AGRICULTURE METEOROLOGY AND WATER MANAGEMENT

AGRICULTURE METEOROLOGY AND WATER MANAGEMENT

Nekesah T. Wafullah and Kuria C. Munene

Delve
PUBLISHING
www.delvepublishing.com

Agriculture Meteorology and Water Management

Nekesah T. Wafullah and Kuria C. Munene

Delve Publishing
224 Shoreacres Road
Burlington, ON L7L 2H2
Canada
www.delvepublishing.com

Email: orders@arclereducation.com

ISBN: 978-1-77469-907-2 (Hardcover)

ABOUT THE AUTHORS

Nekesah T. Wafullah is a skilled agriculture expert from Kenya with extensive knowledge in agricultural energy value addition products, agricultural business management services, project management, various forms of fertilizer, their production, sales, marketing aspects and application regimes; cross border fertilizer trade policies; youth and women empowerment and volunteerism. She is adept at project planning and management as well as creating simple solutions to complex problems. She has experience within agricultural markets in Kenya, Zambia, Malawi, Tanzania, Rwanda, the Democratic Republic of Congo (DRC)- Bukavu and Lubumbashi and Uganda. She mentors high school and college students and advocates for better performance in Agricultural science. During her free time, she loves editing books, watching movies, cooking, baking, networking, reading and dancing. Nekesah holds an M Sc. In Agricultural and Applied Economics degree from the University of Nairobi with a major in International Trade and Policy.

Kuria C. Munene is an agriculture graduate with a masters in Agricultural and applied economics. He has extensive knowledge in agricultural research, policy analysis, supply chains, project management and crop management. He has experience in the use of data analysis tools that led to- high quality inferences for the betterment of the agriculture sector. He is an avid reader and an outdoor sport enthusiast in his spare time.

TABLE OF CONTENTS

LIST OF FIGURES

LIST OF TABLES

LIST OF ACRONYMS

ASTM	American Society for Testing and Materials
AVHRR	Advanced Very-High-Resolution Radiometer
CAgM	Commission for Agricultural Meteorology
CO2	Carbon Dioxide
emf	Electromotive Force
FAO	Food and Agriculture Organization
FAS	Foreign Agricultural Service
GEBA	Global Energy Balance Archive
GIS	Geographical Information Systems
HF	Heat Flux Plates
IBL	Internal Boundary Layer
IEC	International Electrotechnical Commission
IMD	India Meteorological Department
INSAM	International Society for Agricultural Meteorology
IRRI	International Rice Research Institute
ISO	International Organization for Standardization
LAI	Leaf Area Index
LAI	Total Leaf Area
LEISA	Low External Input Sustainable Agriculture
NDC	National Data Centre,
NDVI	Normalized Difference Vegetation Index
PAR	Photosynthetically Active Radiation
PV	Photovoltaics

R.H	Relative Humidity
RTDs	Resistance Temperature Detectors
RTU	Remote Terminal Unit
SCADA	Supervisory Control and Data Acquisition
SMD	Soil Moisture Deficit
SRG	Standard Rain Gauge
SVAT	Soil Vegetation Atmosphere Transfer
TDR	Time Domain Reflectometry
USA	United States of America
USDA	United States Department of Agriculture
UV	Ultraviolet Radiation
VIs	Vegetation Indices
WMO	World Meteorological Organization
WRR	World Radiometric Reference

ACKNOWLEDGMENT

The book is the product of great effort and time spent. The completion of this book could not have been possible without the participation and assistance of so many people whose names may not all be enumerated. Their contributions are sincerely appreciated and gratefully acknowledged. However, we would like to express our deep appreciation and indebtedness, particularly to the following: Arcler Education, Inc. and James M Kihungi for their endless support, kind and understanding spirit during this undertaking. To all relatives, friends and others who in one way or another shared their support, either morally, financially or physically, we would like to personally thank you.

PREFACE

Agrometeorology is the study of weather and the application of climate and weather data to improve or extend agricultural yields or to boost agricultural productivity. Agrometeorology primarily deals with the interactions between meteorological and hydrological elements and agriculture, which includes horticulture, animal husbandry, and forestry. It is an interdisciplinary, comprehensive science that connects the physical and biological sciences to other fields as well. It deals with a complicated system that includes soil, plants, the atmosphere, alternatives for agricultural management, and others. These components interact dynamically on a range of spatial and temporal scales. To produce practical operational applications or recommendations for stakeholders, a thorough understanding of the fully coupled soil-plant-atmosphere system is required. Due to these factors, this book examines the fundamentals of agrometeorology in great detail and demonstrates cause-and-effect relationships and principles that describe the influence of the state of the atmosphere, plants, and soil on various aspects of agricultural production. It also discusses the nature and significance of feedback between these components of the system. It discusses facts and figures from a variety of important essential sciences, including agronomy, hydrology, meteorology, crop and animal physiology, and soil physics and chemistry.

The information in the book is also used to explain complex models that are focused on different system parts like mass balances (such as soil carbon, nutrients, and water), biomass production, crop growth and yield, and crop or pest phenology in order to detect sensitivities or potential responses of the soil-biosphere-atmosphere system. This book illustrates various model constraints and how they might be reduced, even though the agrometeorological data provides common models. The book demonstrates the best ways to operationalize various applications (monitoring, forecasting, warning, suggestions, etc.) at different scales, which is essential for stakeholders. The book also highlights the need for customized goods and information sharing for making both short- and long-term management decisions that are successful. These books discuss solutions for sustainability and improvement (such as risk management, mitigation, and adaptation) while taking into account climatic fluctuation and change.

CHAPTER 1

Introduction to Agrometeorology

CONTENTS

DEFINITION OF AGROMETEOROLOGY

Agrometeorology, which is short for agricultural meteorology, is the study of the logical, climatological, and hydrological conditions that are indicative of agriculture as a result of their interaction with the objects a farmer deals with on a daily basis (Smith). This helps with the wise use of land, speeds up food production, and prevents irreversible abuse of land resources (Molga, 1962). "Biometeorology is an interdisciplinary study concerned with the application of fields of meteorology and climatology to biological systems," according to the definition of the term established by the International Society of Biometeorology (ISB) (Hoppe, 2000). The broad range of interactions between atmospheric processes and living things, such as plants, animals, and people, is included. According to this description, there are basically three subspecialties of biometeorology: plant, human, and agrometeorology (Hoppe, 2000). Agrometeorology's focus is on the plant and animal subbranches. Agrometeorology does not include the third subfield, biometeorology.

Agricultural meteorology's main goal is to increase and completely use our understanding of atmospheric and related phenomena to optimize sustainable agricultural production while making the best use of available meteorological resources and causing the least amount of environmental harm. The quantity and quality of agricultural crops, lumber, and other forest products (like natural rubber), vegetable fibers (like cotton, flax, and sisal), and animal goods and byproducts must all be improved in order to achieve this (e.g., hides). The preservation of natural resources is a second goal. A particular form of land use may be constrained by the climate at a specific location and time. When considering issues related to land use, resource exploitation, and the use of technical processes, an agricultural meteorologist should be contacted (but is frequently not). For instance, the cultivation of semi-arid grassland has frequently (for instance, in the former Soviet Union) been done at the expense of long-term harm from wind and water erosion. The ability to consider processes on a variety of time scales is a benefit of being a meteorologist.

IMPORTANCE AND APPLICATIONS

Agriculture is one of the few human endeavors that is so weather-dependent. Despite the remarkable advancements in agricultural technology over the past fifty years, the majority of agricultural production is still influenced by weather and the environment. Agrometeorological services are more important than ever due to the difficulties presented to various agricultural production methods by rising climate variability and related extreme events, as well as by climate change, all of which have an impact on socioeconomic conditions, particularly in developing countries.

General Importance

Strategic agrometeorological decisions in long-term planning of agricultural systems are guided by knowledge of the environmental resources that are available in time and space as well as the conditions that range from below the soil surface through the soil-air interface into the boundary layer of the atmosphere, both favorable and unfavorable and all varying greatly. Designs for irrigation and drainage systems, decisions for land use and farming practices, and within these decisions, selections of crops and animals, Varieties and breeds, and farm equipment are typical examples. Ecology and economics are on equal footing in modern agriculture, and because of environmental concerns, they are even intertwined. Resources are in short supply, ecological systems are being destroyed, and other environmental problems are getting worse. The widespread and unchecked use of chemical fertilizers and plant protection chemicals is not only bad for the environment, but it also has a significant financial impact on farmers. For tactical agrometeorological decisions in short-term planning of agricultural operations at various growth stages, detailed observations/monitoring, real-time dissemination of meteorological information, quantification by remote sensing (radar and satellites), and derived indices and operational services are important. This information, along with related alerts and services, must be produced and disseminated in a well-organized, automated, and coordinated manner wherever possible. In low external input sustainable agriculture (LEISA), tactical decisions include "average cost"-type judgments on the scheduling of cultural practices such as plowing, sowing/planting, mulching, weeding, thinning, pruning, and harvesting. They also include high-cost choices like applying water, using a lot of chemicals, and using pricey crop protection methods, especially for high input agriculture.

For decision-makers (farmers and managers), regardless of the decision type, an ever-improving understanding of the effects of weather and climate on soils, plants, animals, trees, and related production in farming systems is necessary to use meteorological and climatological information as well as agrometeorological services for agriculture in a timely and efficient manner. To achieve these goals, the proper combination and blending of conventional adaptation tactics, modern scientific and technological knowledge, and suitable policy contexts must be developed. Yields with the existing production means will stay below ideal without policy support mechanisms for agrometeorological services.

Applications

Depending on the time frame, the usefulness of this knowledge depends on the availability and precision of weather and climate forecasts or anticipated weather and climate patterns. The specifications cover everything from precise information for short-term weather forecasts (less than two days), medium-term forecasts (less than ten days) at specific critical periods, to seasonal projections of climatic trends. Indications of potential climatic variability and of increasingly frequent and serious extreme events in the context of global climate change are required as agrometeorological services in addition to the application of other agrometeorological information to ensure that development plans are not rendered meaningless by a significant change in weather and climate behavior. Agrometeorological forecasts may be used to provide important services, such as information on the dates of phonological events, the quantity and quality of crop yields, and the occurrence of animal and crop epidemics, even though reliable long-term weather forecasts pertinent to the agricultural community are not yet regularly available throughout the world. These forecasts rely on known connections between early-stage weather effects and the eventual occurrence, which is anticipated to occur beyond the forecast's release date. This method of "crop prediction without weather forecasting is particularly promising for the assessment of crop conditions in order to be able to identify and statistically analyze probable production anomalies as soon as possible. Both surpluses and deficits are organized or manifest themselves on a national, regional, and global level. Therefore, the consequences of yearly variations in weather patterns as well as possible climatic variabilities and changes on agricultural yields must be considered for long-term planning for the world's food production. There are several elements that affect the climate on a global scale. In the atmosphere, CO2 and water vapor are two of the most crucial elements. In addition to the oceans, trees release water vapor and absorb CO2. Massive amounts of CO2 are released when forests are burned. Therefore, it is essential to encourage regrowth and safeguard forests from various types of devastation, such as that caused by insects, illnesses, and pollutants, as well as by fire and humans. As a part of agrometeorology, forest meteorology offers essential data and services for use by forest managers, foresters, and, in the event of a forest fire, fire brigades. Increasing population pressure and shifting agricultural techniques place a larger burden on agrometeorological services in emerging nations.

Future agricultural communities can anticipate increasing demands for agrometeorological data and services related to technology, farming system patterns, water management, weather-based pest and disease control, etc., preferably with local innovations as starting points. Therefore, one of the problems for the future is the need to prioritize a bottom-up strategy so that predictions, detailed advisories, and contingency planning can be used by even small farmers in their planning and regular agricultural operations. In developed nations, agrometeorological services are primarily concerned with providing national policy and decision makers with environmental data and information in order to support sustained food production, sustainable development, carbon sequestration in agro-ecosystems, and land management techniques that have an impact on the exchange processes of greenhouse gases. Technology transfer may play a role since rich countries may already have or develop technology to first adjust to climate change and climate variability more quickly, but local innovations continue to be crucial for application under the very different conditions in developing countries. Organizations like WMO/CAgM, FAO, and INSAM are already contributing to the development and establishment of agrometeorological services and the transmission of agrometeorological information, and they will be required to do more in the future. Among other things, advisories

i. Information on operational crop protection of all kinds, typical sowing dates, and anticipated sowing dates for the current season, at various temporal and spatial resolutions, are all useful in dry regions.

ii. Data on pest and disease infestations based on weather data and agrometeorological services in a location-specific and user-friendly style are available in more humid areas.

Other important advisory fields that require attention are:

- Management and modification of microclimate.
- Meteorological information for guiding irrigation and drainage.
- Environmental risks and disaster mitigation.
- Highland and mountain agriculture.
- Prediction of El-Nino and rainfall variability data for agricultural planning.
- Information on weather-based pesticides/insecticides applications.

- Arial transport of pollutants and knowledge regarding low level winds for operational activities.
- Workday probabilities (e.g., in marine and lake fishing).
- Agro advisory services for farmers on a regional level to strengthen and provide accurate forecasts and advisories for the farming community.
- Communication of information in a format/language understandable to users.

In more advanced agricultural production, with potential for technology transfer where the absorption capacity exists, we may add:

- Agricultural weather modeling, with a focus on simulation models for crop growth.
- Formulation of intricate data collection methods, as well as quick processing and analysis of sizable spatial data collections.
- The use of geographic information systems (GIS) for crop planning at sizes smaller than those used now.
- The production of information and alerts for wide areas using remote sensing technologies.
- Carbon sequestration measurement.
- Internet and audio-visual media are used to deliver information to users quickly.

Forecasts of important meteorological events that lead to the issuance of advisories and warnings for long enough lead times are of utmost importance. By implementing swift strategic planning, early warnings of natural disasters not only assist in saving the crop but also aid in the advancement or postponement of agricultural operations. Electronic media may play a crucial role in the real-time dissemination of such alerts to end users, which could be crucial for crop protection and productivity.

PHYSICAL CLIMATOLOGY AND AGROMETEOROLOGY

A discipline of meteorology that studies how weather and climate affect crops, rangeland, cattle, and different agricultural enterprises. Micrometeorology, also known as crop micrometeorology for managed vegetative ecosystems and animal biometeorology for livestock operations, is the area of agricultural meteorology that deals with atmospheric-biosphere processes that take place over relatively brief time periods and at small spatial scales. Agricultural climatology is the field that investigates the mechanisms and effects of climatic influences on greater temporal and spatial scales. Agrometeorology, sometimes known as agricultural meteorology, deals with issues that frequently call for knowledge of the biological, physical, and social sciences. The processes it analyzes range from those that take place in the earth, where the deepest plant roots develop, to those that take place in the atmosphere, where seeds, spores, pollen, and insects may be found. Agricultural meteorologists frequently collaborate with researchers from several fields. To comprehend the interactions between vegetation and animals and their atmospheric surroundings, agricultural meteorologists gather and analyze weather and climate data. Making informed decisions about how to best manage the resources used by agriculture, how to maximize agricultural production, and how to adopt farming practices to reduce any unfavorable effects of agriculture on the environment all benefit from the climatic data produced by agricultural meteorologists. Such knowledge is essential to maintaining agriculture's economic viability and environmental sustainability both now and in the future. Additionally, agricultural meteorologists collect, assess, and share data on the effects and outcomes of climate variability and change on agriculture.

Agricultural meteorologists are increasingly helping policy makers create plans to deal with climatic calamities like floods, hail, or droughts as well as climatic developments like global warming and climate variability. Agronomic and horticultural crops, trees, and animal production, as well as the final transportation of agricultural goods to markets, are all facets of agriculture that agricultural meteorologists are involved in. On geographical scales ranging from a leaf to a watershed and even a continent, they research the energy and mass exchange processes of heat, carbon dioxide, water vapor, and trace gases like methane, nitrous oxide, and ammonia within the biosphere. For instance, they research the photosynthesis, output, and water use of particular leaves, entire plants, and entire fields. Additionally, they look at climatic processes over timescales of less than a second to more than a decade.

Agricultural Climatology

The general study of climate and how it affects crops; for instance, the relationship between growth rate and crop yields and numerous climatic elements and, as a result, the best and worst climates for any particular crop. called agroclimatology as well.

Biometeorology

A field of meteorology and ecology that studies how climate change affects humans, animals, and plants. Maintaining adequate thermal homeostasis with their environment is the main challenge for living things. Natural mechanisms exist for organisms to withstand challenging circumstances. These strategies include acclimation, dormancy, and hibernation, or an organism may choose to relocate to a better environment or microenvironment in specific circumstances. Humans frequently employ technology to create a welcoming environment.

Climatology

The examination of climate by science. Climate is the average and range of predicted weather conditions for a specific area, time of year, and season. The mean values of meteorological variables, such as temperature, precipitation, wind, humidity, and cloud cover, are frequently used to characterize the climate. A thorough explanation also takes into account these quantities' high values and variability. The weather in a place frequently varies seasonally and throughout the day, with January typically having a considerably different climate than July. On a regional and global scale, the climate also shows notable year-to-year variability and longer-term variations. Its objectives are to provide a thorough description of the Earth's climate on a variety of scales, to comprehend its characteristics in terms of fundamental physical laws, and to create models of the climate for sensitivity studies and the forecasting of future changes that could be brought on by both natural and human causes.

Crop Micrometeorology

The area of meteorology concerned with how crops interact with their immediate physical environment.

Micrometeorology

The investigation of micrometeorological phenomena connected to the interaction between the atmosphere and the surface of the Earth. Interactions taking place in the lowest atmospheric layers determine the lower boundary condition for the atmosphere and the higher boundary condition for the subsurface soil or water. Turbulence in the atmospheric boundary layer transports energy, heat, water vapor, different gases, and particulate matter vertically, creating the habitat for plants and animals at the surface. These interactions are crucial for supplying the atmosphere with energy and water vapor, which in turn controls broad-scale weather and climate patterns. Micrometeorology also studies the deposition of pollutants near the surface as well as how air pollutants are dispersed and carried through the boundary layer. Atmospheric motions with time periods between 15 minutes and one hour are frequently very modest. This spectral gap justifies the distinction between micrometeorology and other branches of meteorology. Time scales of less than 15 minutes to an hour and horizontal length scales of less than 2 to 10 kilometers are studied in micrometeorology, which is a branch of meteorology. Dust devils, mirages, the creation of dew and frost, evaporation, and cloud streets are some of the phenomena that micrometeorology studies.

Ecosystem

An entire community of living things and the nonliving components of their environment make up an ecosystem. Soil, rocks, minerals, plants, animals, and microorganisms are among its constituents, as are nearby water sources and the local atmosphere. Ecosystem sizes are quite variable. An ecosystem can range in size from a puddle to a family garden to an enormous rainforest that spans a region greater than many countries. Given the abundance of microorganisms found there, even an animal's body could be regarded as an ecosystem. On a far broader scale, the development of different human societies offers an illuminating example of how ecosystems have impacted civilizations.

Weather Observations

The gathering, storing, and sending of information about the weather and its variable components. For example, in India, the collecting of meteorological data is one of the main duties of the National Data Centre (NDC), a component of the India Meteorological Department (IMD). The data are transmitted to the IMD NDC through a variety of communication techniques. The Center's huge computers are programmed to plot, analyze, and process the raw data as well as create predictive weather charts from it. Field offices, other governmental organizations, and other stakeholders are then informed of the processed data and the forecast information via specialized National Weather Service technologies and regular telecommunications. Based on both processed and unprocessed data, they then generate forecasts and warnings. To meet the demands of meteorologists, climatologists, and users in forestry, agriculture, aviation, and other industries, a wide range of meteorological data is necessary.

Weather Forecasting and Prediction

These include methods for gathering and analyzing meteorological observations to create and distribute information about upcoming weather conditions. According to the expected phenomenon's spatial and temporal scale, weather forecasts can be categorized. Turbulent atmospheric fluctuations are those having a period of less than 100 s and a length of less than 100 m (330 ft). Micrometeorology, the study of atmospheric turbulence, is crucial for comprehending the dispersion of air pollutants and other features of the climate close to the ground. Typical meteorological measurements are performed using sampling strategies that eliminate turbulence's influence. The terms mesoscale, synoptic scale, and planetary scale are used to describe three categories of events that occur on scales larger than the turbulent microscale. All moist convection events, from individual cloud cells to convective cloud complexes linked to tropical storms, prefrontal squall lines, and the intertropical convergence zone, are included at the mesoscale. The sea wind, mountain valley circulations, and the intricate structure of frontal inversions are also categorized as mesoscale phenomena. Most mesoscale phenomena occur over shorter timescales than 12 hours. There is ongoing research into the prediction of mesoscale phenomena.

The majority of forecasting techniques rely on empirical laws or the short-range extrapolation of recent observations, especially those made possible by radar and geostationary satellites. To represent the erratic nature of the events, forecasts are typically expressed in probabilistic terms. Given the considerable hazards to life and property posed by many mesoscale occurrences, it is customary to issue advisories of likely occurrence far in advance. These "watch" advisories encourage the general people to get as prepared as the possible hazard warrants. The advice is changed to a "warning" when it is determined that the phenomena is about to occur, with the hope that the public would act quickly to stop the loss of life. The synoptic scale, which refers to the next-largest scale of weather events, is defined by the network of meteorological stations that provide simultaneous observations, or synoptic observations. Synoptic-scale phenomena include the migratory storm systems of the extratropics as well as the accompanying undulating upper-air circulation wind currents. Barometric minima, often known as lows, depressions, or cyclones, are linked to the storms. The synoptic method of forecasting involves gathering weather data at the same time as charting and analyzing it on a map of the world. An expert analyst can track the development and intensity of weather systems and predict their positions after studying several of these maps in chronological order. Large networks of data must be regularly and frequently used in this forecasting technique. Planetary-scale events are long-lasting, essentially stationary perturbations of the Earth's atmosphere with dimensions on the horizontal that are similar to its radius.

The principal orographic features of the world and the latent and sensible heat sources provided by the oceans appear to be associated with these general circulation's dominant characteristics. They frequently direct the trajectories taken by synoptic-scale storms and use synoptic transients as a further source of energy and momentum. The prediction of weather events using numerical solutions to the equations regulating the mobility and changes in the state of the atmosphere is known as numerical weather prediction. In addition to being utilized for short-range weather forecasting, numerical weather prediction techniques are also used in studies on the movement of air pollutants and the impact of greenhouse gases on global climate change. Since there was only one layer in the first operational numerical weather prediction model, it could only simulate the temporal variation of the mean vertical structure of the atmosphere. Nowadays, computers make it possible to create multilayer models (often 10–20 levels deep) that might explain the vertical change of the wind, temperature, and moisture. For vast scales of motion, these multilevel models forecast the fundamental meteorological variables. Weather agencies in many nations employ global models with horizontal resolutions as fine as 125 mi (200 km). The most powerful computers are needed to create a 10-day forecast using global numerical weather prediction models. Global research models may be used for climate studies if they were run for considerably longer periods of time. A more accurate numerical representation of the energy transfer and turbulent dissipative processes within the atmosphere and at the air–earth boundary, as well as significantly increased computing machine speeds and capacities, are required for the extension of numerical predictions to long time intervals (many years). The mean circulations produced by long-term simulations of climate models closely reflect the atmospheric circulation. Even though it is hard to anticipate the daily variations of weather for very long periods, these simulations have proved helpful in understanding the main characteristics of the Earth's climate. Climate models are currently being used to forecast future changes in the climate brought on by changes in the atmospheric composition or characteristics of the Earth's surface because of human activities. Climate models have also been successfully used to explain paleoclimatic fluctuations. A wide continental data network is regularly used to gather surface meteorological observations, with most of these observations coming from both hemispheres' central latitudes. Similar in-place observations from oceanic locations are provided by military vessels, fixed and floating buoys, and commercial ships of opportunity. It is customary to gather data on the winds, pressure, temperature, and moisture across the troposphere and into the stratosphere;

- direct samples of the free atmosphere taken by commercial and military aircraft and balloon-borne sensor packages (radiosonde observations);
- a network of Doppler radars operated by the National Weather Service, wind profilers (vertically oriented Doppler radars), and lidars; and
- Onboard geostationary or polar orbiting spacecraft, specialized sensors are used.

In data-scarce (mainly oceanic) portions of the earth, the remotely sensed observations from meteorological satellites have been extremely helpful in giving essential measurements of a truly and vertically averaged temperature, moisture, and winds. Such measurements are required to support contemporary numerical weather prediction techniques and to allow forecasters to continuously monitor the activities of major storms (like hurricanes). Longer term (greater than two weeks) and shorter-term categories are used to categorize forecast products and forecasting ability. These different levels of proficiency reflect the fact that while current numerical prediction models, such the medium-range forecast, have gotten quite excellent at forecasting large-scale circulation and temperature, they have had less success in forecasting weather. Predicting precipitation amount and kind based on the presence of precipitation and convection is one example. Because mesoscale processes are becoming increasingly crucial to the forecast's overall skill, each of these predictions becomes progressively more challenging. A relatively short-range weather forecasting technique is casting. The phrase "now casting" is occasionally used in a broad sense to describe any region-specific forecast for the period up to 12 hours in the future that is based on incredibly thorough observational data. However, a narrower definition of nowcasting would likely be the full description of the current weather conditions together with forecasts generated by extrapolation up to around two hours in the future. In many instances, it is possible to produce useful extrapolation forecasts for longer periods of time, however, in some weather scenarios, the accuracy of extrapolation forecasts decreases quickly over time due to the development or decay of the weather systems. Long-range or extended-range climate predictions are forecasts of time averages of atmospheric variables, such as sea surface temperature, where the lead time for the prediction is greater than two weeks. Climate outlooks are long-term forecasts of monthly and seasonal average temperatures and precipitation. Long-range outlooks have never been particularly accurate because the predictions must consider a wide range of potential outcomes, while the observed single event against which the outlook is verified includes the noise produced by the particular synoptic disturbances that actually occur and are unpredictable on monthly and seasonal time scales. Some estimates of potential predictability indicate that in middle latitudes, the noise is typically greater than the signal.

AGROMETEOROLOGY SCOPE

Further to what is said in the section above, other simplified approaches to understanding the scope of agrometeorology are provided as follows.

Soil and water: The process of weathering plays a crucial role in the formation and subsequent characterization of a soil type. The mechanical, physical, and chemical characteristics of the soil, the creatures it supports, and its capacity to store and release heat, nutrients, and moisture are all impacted by current and past weather conditions. Rainfall not only enriches the soil chemically but also removes (or "leaches") soil nutrients. Local weather conditions have a significant impact on the mechanical status of the soil, which impacts pasture management (stocking density, etc.), crop cultivation, insect control, and harvesting.

Environmental scientist cooperation is crucial in these situations; for example, meteorologists must avoid oversimplifying the surface elements of their work, while hydrologists must avoid presuming to be fully knowledgeable about all weather and climate phenomena. Overall, there is a need for improvement in the collaboration of meteorologists, hydrologists, and soil scientists. The soil serves as a water storage area for plant growth and development. The effectiveness of irrigation or precipitation on the one hand, as well as the physical characteristics and depth of the soil on the other, determine how much water is present in the soil. The climate, the physical characteristics of the soil, and the plant community's root system all affect how quickly water evaporates from the soil. Agrometeorology is very concerned with the efficient use of soil water and cultivation methods for maintaining soil moisture, especially in semi-arid regions. Regional and local weather conditions affect erosion caused by wind and water. Additionally, the presence and strength of the vegetative cover affect how much a given piece of ground is eroded by erosive agents. In places with substantial evaporation and inadequate drainage, especially when irrigation water is of low quality, there may be an excessive concentration of salt at the soil surface.

The start and end of the rainy season in (sub)tropical countries; the time when the water content of the upper soil layers exceeds the "field capacity" in wet climates; or the length of the frozen ground period in high latitudes are all examples of how seasonal changes affect soil conditions and, consequently, the farming program (or "calendar").

Plants and crop microclimate: Environmental factors affect a plant's growth and development at every stage (Woodward, 1987). Understanding how the environment's structure (ground cover, surface slope, degree of shelter, etc.) and the local microclimate, in the crop and around the crop, are related may lead to activities intended to improve the growth condition over the long run. The influence of the weather should be considered even before planting. The weather during the year that the seed was generated determines the quality of the seed that is sown. The weather over many prior seasons can also have an impact on the productivity of long-rotation crops, such as vines, fruit, and forest trees. Seasonal weather has an impact on post-harvest activities including drying grass and other crops and the ability to preserve the quality of stored farm products. Knowing the weather and climate is crucial for preventing forest, bush, and grass fires, as they play a significant role in their occurrence.

Farm animals: Animals are impacted by the weather at all phases of development and health, whether they are well-fed and healthy or malnourished and ill. It primarily affects their nutrition, growth, health, and ability to produce offspring directly, such as through drought or the impact of excessively high or low temperatures (Mount, 1979). Additionally, the state of the soils where animals are housed and their access to food (grass and crops) are both impacted by weather. Animal geographical distributions are influenced by climate. Weather affects the quantity and quality of animal products as well as their processing, storage, and transportation. Keane (1986) goes into detail about a variety of weather-related effects on animals.

Diseases and pests: Weather has an impact on how vulnerable plants and animals are to pest attacks and diseases, as well as how likely they are to host them. Additionally, it influences the biology of pests and disease-causing organisms, as well as the types, quantities, and behaviors of pests (as well as those of their predators), as well as the severity and virulence of diseases. The efficiency of applied control or eradication strategies, as well as the proliferation and aerial transport of pests and illnesses, are dependent upon atmospheric agencies in the protection of crops and cattle. Examples of this include the wind-dependent dispersal of chemical sprays, the impact of temperature on the chemical activity of pesticides or on the susceptibility of pests to such treatments, and the fungi's sensitivity to the presence or absence of liquid water, whether it is dew or rainfall.

Farm buildings, equipment and operations: Climatic factors must be taken into consideration while designing farm structures, especially animal housing and crop storage facilities. The use of manpower, the method of harvesting, and the selection of farm equipment are all climate-sensitive decisions (Dalton, 1974). The climatological likelihood that the soil will be robust enough to sustain machinery when it is needed is an important consideration. Even plans for a single day call for predictions for several days, and those forecasts must include some factors not typically included in public forecasts. Relevant forecasts that are made far enough in advance determine the farmer's best choice of work priorities.

Artificial modification of meteorological regimes: Aspects of the local environment like soil moisture, wind speed, and atmospheric humidity are significantly influenced by irrigation, windbreaks, shelterbelts, mulch, snow and water storage, conservation, and soil cultural practices. The utilization of glasshouses and specialized animal housing allows for the maximum degree of environmental control. It is possible to develop various levels of independence from external environmental factors (at times to a very high degree). However, when the gap widens between the intended internal climate conditions and the actual weather and climate outside the structure, the cost of the materials and the energy required to attain this independence tends to rise. The cost-benefit ratios of these artificial climates are kept within an acceptable range by using meteorological support in their design and administration.

Climate change: Numerous studies have been conducted on the agricultural impacts of climate change caused by or facilitated by human activity, both qualitatively and quantitatively (e.g., changes in carbon dioxide content of the atmosphere due to the consumption of fossil fuels, such as coal and oil, or the production of methane by animals and vegetation). For several reasons, this section will not go into detail with the issue of climate change. The first is that since climate change is probably going to happen gradually, agriculture may frequently be able to adjust to environmental changes that are understood in advance. The second is that there is still a lot of study being done on processes associated with climate change, thus it is possible that in a few years, clear assertions about their effects may need to be reevaluated.

Changes in the global temperature can cause climate changes at regional or local scales that vary in sign and extent. The main justification for not treating the topic of "climate change" separately here is because it is complicated when considered in the context of agrometeorology. Numerous researchers focus solely on the issue of the influence of rising average temperatures (during the day, at night, or continuously? They frequently appear unaware that such an increase may involve numerous additional weather changes that are not taken into account in their model (e.g., seasonal or annual?) on crop growth. For instance, greater temperatures may result in increased evaporation (although this is not necessarily the case), as well as a probable rise in aridity or cloudiness.

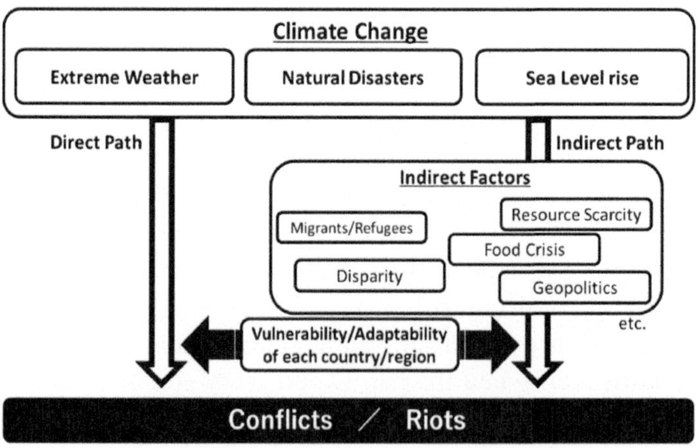

Figure 1: *The effect of various factors.*

Source: Sekiyama T. Climate Security and Its Implications for East Asia. Climate. 2022; 10(7):104. https://doi.org/10.3390/cli10070104Successes rarely receive as much attention as disasters, yet many potential catastrophes can, to a large part, be averted or at the very least, have their effects lessened by human intervention. It is intriguing to learn that every indicator is weather-dependent!

Fighting these harmful weather effects is a major focus of agrometeorology. Understanding how weather and climate affect vegetation and cattle, however, can sometimes have wholly positive consequences that boost agricultural activities in unanticipated ways.

The latter would reduce incoming radiation and maybe (or possibly not) enhance precipitation. Increased photosynthesis would result from higher temperatures and enough radiation, but shorter growth seasons could result from concurrently increased metabolism, which could result in a net drop in productivity. Early high temperatures in the spring at temperate latitudes could result in early flowering at a time when the risk of frost is still quite high. Such climatic adverse effects are numerous (Fajer and Bazzaz, 1992). In a nutshell, climate change is an unsuitable topic for an introductory agrometeorological text due to its complexity. To avoid unwarranted simplifications of the weather or biological processes, agronomists and professional meteorologists should cooperatively examine climate change issues. Variability should be seen as an equally essential topic as potential change.

AGROMETEOROLOGICAL INFORMATION

We shall mention a few of the user types that meteorologists are expected to service and demonstrate how their demands differ in order to showcase the range of information requirements in agrometeorology. These user groups are not listed in any particular priority order, and there may be regional variations in their traits, circumstances, and demands. Owners of big farms sometimes need to make significant expenditures in mechanization and occasionally also in (part-time) manpower. In general, they cultivate a small variety of crops, sometimes developing into monoculture. Such agricultural factories may need substantial meteorological and climatic information for planning, sometimes in a computer-compatible format, to operate properly. When dealing with high-value crops cultivated over a very small area, a horticulturist will frequently try to regulate the climate as much as possible in a greenhouse or in plastic tunnels. He must regularly have access to quantitative information, such as the local humidity and cloudiness, in order to make the appropriate modifications to control the radiation traveling through his roof and the ventilation air coming below it. Family farms fall somewhere in the middle of the two aforementioned market-directed farming models in terms of scale. Their primary exports are typically livestock or local mainstays like grapes, grains, beans, beets, fruits, sugarcane, and vegetables, along with a few ancillary minor crops or animals. These farmers frequently follow agro-weather broadcasts, which help them make short-term operational choices (especially on risks like frosts). For example, cattle farmers need to know when they can make hay in order to provide their herds with winter forage.

Farmers in regions without winter are interested in early warnings of coming catastrophes like floods or storms. To provide an "early warning" of crop failures, government planners (from, say, the Ministry of Agriculture) may be tasked with predicting the yield of the upcoming national staple crop harvest.

This implies that the meteorologist must be knowledgeable about the validity of current crop growth models that account for weather. The climatological suitability of diverse places for cultivating new types of crops is another area for government advice.

If they are able to employ predictions couched in terms of probability, agricultural product manufacturers and other related businesses involved in processing or shipping food for the market can use long-term forecasts and meteorological information for planning purposes.

Statistical climate research models are preferred over climate models (Changnon, 1992). Researchers with expertise in related fields like hydrology and agronomy frequently request meteorological datasets. Use this to kick off collaboration.

AGROMETEOROLOGICAL SERVICES

It is vital to establish an agrometeorological unit in a National Meteorological Service where the necessary infrastructure for data collecting and analysis is already in place along with forecasting services and a national database in order to appropriately handle such a wide range of demands. Following the foundation of such a unit, the following three stages of development were suggested by a WMO CAgM working group in 1980:

- A "passive" pre-operational stage that makes use of the infrastructure already in place and more effectively provides basic climatic data for agriculture by delegating this responsibility to a select group of employees of a climatological service. At this time, local weather-sensitive issues should be identified by a national agrometeorological committee with assistance from WMO and FAO.

- A more "active" stage during which a specialized unit offers derived information (such as increasing degree days and leaf-wetness duration). Some nations might make a stop here.
- A stage of operation where forecasts and advisory services are offered, and the essential research is done to provide such problem-focused services.

The tactics and strategy of food production are two kinds of issues that agrometeorology and agroclimatology may help with. Crop selection, animal selection, the size and scope of farm equipment, and soil cultivation techniques are all strategic concerns. The appropriate selection of land usage, production facilities, and genetic issues is part of strategic planning. A statistical analysis of previous climates, their variability, and the spectrum of potential outcomes, notably extremes and sequences, which the atmosphere can produce, can also be helpful in determining the risk of proposed future actions. The short-term scheduling of agricultural operations and everything else that goes into good husbandry are tactical items. Weather forecasts can be used as a tactical tool, but only if they are accurate and easily accessible, include farming-related weather characteristics, and make explicit the consequences of incoming weather on agriculture so that farmers who are weather-savvy can prepare accordingly.

A forecast is only meaningful if the recipient has a course of action, they can take to either benefit from favorable circumstances or lessen negative impacts. Farmers are frequently helpless to intervene; whereas an aircraft can alter course to avoid a hailstorm, farmers are unable to protect their crops. Understanding the consequences of physical conditions is a requirement for solving all problems.

More understanding of the basic biological processes is necessary for complete comprehension. Accordingly, groups of scientists must do basic research. Most meteorologists are aware that they lack the training necessary to completely understand biology.

However, because they deal with the weather on a daily basis, biologists and agronomists frequently believe they are knowledgeable enough to handle it. It takes judgment and perseverance to approach biological "weather" models with simplifications that are realistic. Interdisciplinary science makes sluggish progress, but one positive aspect of agrometeorology is that valuable work can be done to acquire an interim degree of precision that is helpful and worthwhile, without delaying implementation until the full results of fundamental research are known.

FUNDAMENTAL PRINCIPLES OF AGROMETEOROLOGY

Agricultural meteorology has applicability to the plant, animal, and soil habitats and is based on a foundation of fundamental physical laws.

Radiation: According to the laws of radiation, radiant energy from the sun and the atmosphere is diffused and made available to plants for photosynthesis as well as transformed into various kinds of energy by solid and liquid surfaces. The visible wavelength ranges from ~400 nm to ~700 nm, often known as the shortwave band or, for biological purposes, the band of photosynthetically active radiation (PAR) makes up around 40% of the light the Sun emits. UV-A radiation has wavelengths between 320 and 400 nm and UV-B radiation has wavelengths between 280-315 nm. These wavelengths are just a little bit shorter than those in the PAR zone. Although plants do not utilize ultraviolet radiation for photosynthesis, it can harm living tissue in both plants and animals, especially in basic organisms. A leaf surface may receive visible light directly from the Sun or indirectly through the reflection of clouds, air molecules, and solid or liquid surfaces (including other plant leaves). The amount of energy received varies during the day due to leaf orientations and the solar zenith angle. The term "leaf area index" is used to characterize the portion of a plant that is available for photosynthesis. It compares the plan area of leaves to the ground surface covered by the plant (including only one side of the leaf). The band from around 600–100 nm is known as the near-infrared area, which makes up about 40% of the solar spectrum. Longwave or infrared radiation is defined as radiation with wavelengths just bigger than visible light.

Heat Balance: The ideal gas law and relationships between atmospheric pressure, temperature, and density are based on thermodynamic principles, as is the transfer and transformation of energy (first law of thermodynamics). The energy balance (conservation of energy) for the system being examined, such as a metabolizing organism or a plant-covered or soil surface, is a main emphasis of agricultural meteorology. A portion of this net incoming radiant energy's shortwave component is used by a plant for photosynthesis. When conditions are balanced, the plant employs its ability to evapotranspiration to control its internal temperature by converting extra sensible heat to latent heat. The majority of domesticated animals, including humans, have the ability to expel extra heat through evaporation.

Surface Aerodynamics: Understanding how plants exchange moisture, trace gases, and thermal energy with and extract energy from the free atmosphere through turbulent processes is based on the aerodynamics of their interactions with the atmosphere. Additionally, descriptions of the mean and dynamic flow processes on ranges ranging from centimeters to hundreds of kilometers are necessary for the transport of pollen, spores, insects, and chemical sprays. The more complex profiles seen inside crop canopies are typically defined by empirical relations.

Evaporation and Precipitation: On a considerable portion of the planet, agriculture is conducted when either a surplus or a shortage of water is a significant barrier to producing effective crops. Therefore, the study of precipitation and evaporation is a key area of interest in agricultural meteorology and climatology. When other parts of the heat budget are known, the heat balance equation can be used to predict the rate of evaporation for a surface. Agriculture depends heavily on the quantity, timing, and evaporation of precipitation and water. With growing competition for fresh water due to increased population and expanded uses of water, irrigation scheduling requires trustworthy climatic information as well as good weather forecasts. The need of improving seasonal to interannual forecasts of weather and climate is increased by the vulnerability of many developing nations with little precipitation to interannual variability of precipitation.

Instrumentation, Measurements, and Networks: Agricultural climatology is based on records of fundamental meteorological data made over vast areas and extraordinarily lengthy time frames. These data serve as the foundation for understanding both climate variability and change as well as for deriving statistically meaningful connections between meteorological variables and soil and plant processes, plant, animal, and pest development, and seasonal yield. Agriculturists need measures of soil moisture and temperature in addition to conventional air measurements. These measurements are less frequently taken, despite the fact that their importance in climate memory and consequently seasonal forecasting, particularly with regard to soil moisture, is now being acknowledged. A greater number of these measures and measuring networks are required, especially in underdeveloped nations where the use of technology to lessen vulnerability to climatic variability is severely constrained.

Modeling and Theory: Global climate modelers looking for more accurate representations of energy, momentum, and moisture budgets at the Earth's surface and crop modelers looking for ways to comprehend plant responses to climate and project crop yields are at least two sources from which modeling of plant interactions with the atmosphere has emerged. To the so-called soil-vegetation-atmosphere transfer (SVAT) models, climate scientists contribute meteorological data at each surface grid point at each model time step to create what are known as "surface packages" (a few minutes to hours). The SVAT model then determines how the soil and plants will react (such as through evaporation or transpiration, temperature change, soil moisture content, moisture absorption by roots, rain or dew held on leaves, precipitation runoff, and momentum extracted), and it feeds back to the climate model surface fluxes of heat, moisture, and momentum that are consistent with these soil- and plant-based changes. The level of precision that can be expressed by computational limitations is limited, but even if the models are crude, they offer a conceptual foundation for the ultimate coupling of more precise agricultural, forest, and ecosystem models.

Manipulating Microclimates: Agriculturists have a long history of enhancing crop growth by manipulating soil and plant microclimates, through the use of irrigation, glasshouses, shelterbelts, windbreaks, snow fences, wind machines, surface mulches, certain tillage practices, alley cropping, and agroforestry. The design and operation of such modifications require considerable information on the means, extremes, and interannual variability of climate at the specific location where the practice is implemented. Horticulture crops, which typically have a higher value per unit area than grain crops, are sensitive to small changes in microclimate. Also, in contrast to grains, horticulture crops are more sensitive to weather-induced reduction in product quality or market value. For instance, the desirable red coloration on some fruits is sensitive to optimal amounts of solar radiation at a critical stage. Manipulation of microclimates for horticultural crops is more cost-effective than for cereals because of both their high value and their sensitivity of quality to microclimate. Weather extremes may have multiyear impacts on agricultural crops grown as perennials (e.g., fruits, nuts, and grapes), which raises the cost-effectiveness of microclimate modifications for reducing such extremes.

CHAPTER

2

Radiation

CONTENTS

OVERVIEW

Agricultural meteorology largely focuses on the atmospheric conditions at the lowest layers, where the conversion of solar energy into chemical energy on the surface of the soil and plants most directly affects meteorological events. Because of this, solar radiation has a specific significance in agricultural meteorology as a meteorological parameter.

The soil surface only gets about 43% of the total solar radiation, which equates to an average of 1.94 Cal-cm^{-2}-min^{-1} (solar constant) in vertical impingement on the outer limits of the atmosphere. From the edge of the atmosphere, 42% are directly reflected into space. Air molecules, water droplets, and dust particles in the atmosphere absorb or disperse about 15% of the energy.

The extinction of solar radiation in the atmosphere is the total of absorption and scattering. The route lengths of the sun rays in the atmosphere, the state of the atmosphere, its moisture content, and its dust-turbidity all affect how much energy is lost by absorption and scattering. Lambert's Law, which holds true for all radiation filters, can be used to calculate the relevant permeability of the atmosphere. As a result, the sunlight's vertical irradiation intensity is:

$$I_m = I_o * q_m$$

at the earth's surface, where I_o represents the solar constant, q, the transmission factor for the layer thickness I (solar angle 90°) and m represents the thickness of the amount of traversed air. If, instead of the transmission factor q, one introduces the extinction coefficient a = lnq, the equation is transformed into

$$I_m = I_o * e^{-a.m}$$

The thickness of the amount of air trans-radiated is given approximately by the air pressure b and the solar height h, as follows:

$$m = \frac{b}{760} \cdot \frac{1}{sinh}$$

The values of 'm' at normal air pressure (b = 760 mm) for several solar angles can be seen in the table below. Water vapor, 0_3, and 0_2 are three gases that play a specific role in the absorption of solar energy. The measure for the amount of water vapor is introduced as the height w (in cm) of a liquid water column that would hypothetically exist following condensation of the water vapor (precipitable water).

Table 1: *Some components used in estimating radiation*

Thickness of the air amount (m) in its dependence on solar height (solar angle) (h)								
h	5°	10°	15°	20°	30°	45°	60	90°
Elevation above sea level =0m	10.40	5.60	3.82	2.90	2.00	1.41	1.15	1.00
Elevation above sea level = 500 m	9.80	5.28	3.60	2.73	1.88	1.34	1.08	0.94
Transmission factors and extinction coefficients for total radiation in their dependence on the absolute air-mass for pure, dry air								
Absolute air mass Mp	0.5	1	2	3	4	6	8	10
Transmission factor q	0.900	0.905	0.914	0.921	0.927	0.935	0.941	0.946
Extinction efficient a = Inq	0.105	0.099	0.089	0.082	0.076	0.0671	0.0600	0.055
Water vapor (precipitable water); corresponds to a partial pressure e mm Hg of water vapor								
(mm)	2.0	5.5	7.3	8.5	9.4	11.0	12.0	
w (cm)	0.1	0.5	1.0	1.5	2.0	3.0	4.0	
Weakening of direct solar radiation by water vapor, ozone, and oxygen at various solar angles								
Solar angles:	90° °	60°	30	10°	5°			
H_20(w= 1cm)	6.8%	7.3%	8.7%	12.4%	14.8%			
0_3 (2 mm) and O_2	2.1%	2.1%	2.8%	4.5%	5.7%			

Source: Conceptualized from; Seemann et al., (1979: 2012).

The table above also shows how water vapor, ozone, and oxygen at different sun angles reduce direct solar energy.

Aerosol is the term used to describe the turbidity that dust and other airborne pollutants generate. In addition to contributing to apparent turbidity, aerosols reduce the strength of solar radiation when it enters the atmosphere. The primary metric for turbidity is Linke's turbidity factor. Its value depends on the surrounding area. Compared to winter, it is higher in the summer. According to Linke (1953), the turbidity factor typically has the following values:

- High mountains T= 1.90
- Flat land T=2.75
- Major city T=3.75
- Industrial area T=5.00.

The soil surface receives solar radiation both as parallel radiation, also known as direct radiation and as diffuse radiation due to the impact of the atmosphere (also called sky radiation).

MEASURING SYSTEMS

Gram-calories per square centimeter per minute (Cal cm^{-2} min^{-1}) is a common unit of measurement for the solar energy that is radiated per unit time to a given surface. In recent years, the joule has replaced calories as the primary measurement unit for radiation. Watt is also used occasionally, particularly in publications with technical content. These three units of measurement are compared in the table below. It is helpful to keep in mind that a kilowatt hour is equivalent to 860 kcal for future conversions, among other things.

Table 2: *Measuring units of solar energy*

	$Cal \cdot cm^{-2} \cdot min^{-1}$	$J \cdot cm^{-2} \cdot min^{-1}$	$W \cdot cm^{-2}$
$Cal \cdot cm^{-2} \cdot min^{-1}$	1	4.1868	0.069
$J \cdot cm^{-2} \cdot min^{-1}$	0.238	1	0.00165
$W \cdot cm^{-2}$	14.3	60.6	1

Source: Seemann et al., (1979:2012).

GLOBAL RADIATION

The total amount of solar and ozone radiation on a horizontal surface is known as global radiation. Each year, the entire surface of the earth receives 860 x 1018 kcal of radiation. This is a million times more energy than what the world's industrialized nations produce in electricity combined (Schulze, 1970). Due to the wide regional variation in the many variables that affect sky radiation's strength, a thorough depiction of the phenomenon's global distribution will inevitably remain challenging. When clouding effects are included, this becomes particularly challenging. According to Schulze (1970), there is an average 6/10 cloud cover, which reduces global radiation by around 64 percent.

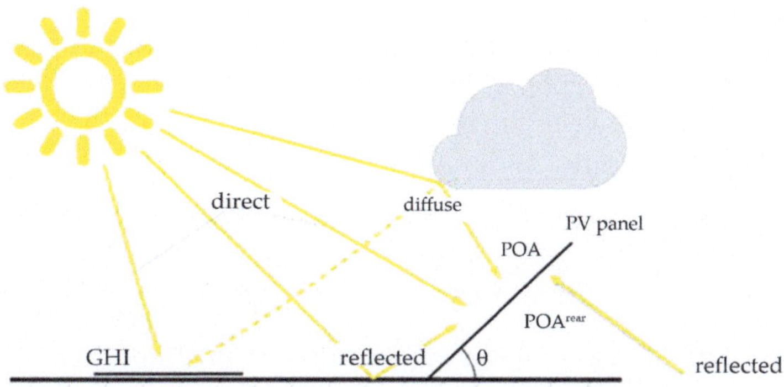

***Figure 2:** Origin of the three types of radiation.*

Source: A.N. Korevaar, M. (2022). Measuring Solar Irradiance for Photovoltaics. IntechOpen. doi: 10.5772/intechopen.105580

For ordinary latitudes, Nicolet and Doginaux (1951) explored the correlations between sunshine duration and radiation sums and developed the formula shown below:

$$s = s_o \frac{s^n}{s_o}$$

where S = total radiation with corresponding clouding; S_o = sum of radiation under cloud-free sky; s= duration of sunshine with corresponding clouding; So = duration of sunshine under cloud-free sky; n = exponent:

$$1 + 0.8\frac{s}{s_o}$$

Radiation on Inclined Surfaces

The previous method of handling solar energy was on a horizontal, flat surface to serve as the receiving surface. The radiation conditions on flat surfaces will only have a very small impact on topo-climatological work; inclined surfaces, or slopes with different angles of inclination and exposure, will be even more affected. Due to the rarity of radiation observations for these, extensive calculations are required. The effects of radiation, however, are particularly significant in topo-climatology, hence this issue will be thoroughly covered here. If the intensity I_m of solar radiation, i.e., its surface strength on vertical incidence at a given time in a location on earth is known, then the strength of irradiation I on any desired surface that forms the angle y (incidence angle) with the solar radiation is calculated as;

$$I = I_m \, siny$$

Here, I_m refers to the plane. The solar angle is then exactly defined as the angle of incidence between the incident sun ray and the surface it impinges. Thus, it is dependent only on the geographic latitude φ, the solar declination δ corresponding to the calendar date and the hour-angle t that results from the time of day. Then:

$$sin\, y = sin\, h = cos\ \phi\ cos\ \delta\ cos\, t + sin\ \phi\ sin\, \delta$$

φ represents the geographic latitude of the location. δ can be derived from the astronomical annuals. The solar vertical's junction with the horizon yields the hour-angle, t. The hour angle is measured as the sun moves across the sky from the daily solar culmination in the south (noon = 0°) favorably to the west (1800 = 90°) to the north (2400 = 180°), totaling 15° for each full hour of the day. Kaempfert (1942) has derived the following formula for the angle of incidence on a vertical wall (Y_w):

$$sin\, y_w = sin\, á\, cos\, \mathrm{h}$$

where á is the "wall azimuth", namely the angle between the wall that would be produced by raising the slope to the vertical, and the solar vertical, counted positively from the South = 0° through West towards North. We get cosh from the relationship for the determination of the solar azimuth: a

$$sin\ a = \frac{cos\ \delta}{cos\ h}\ sin\ t$$

$$cos\ h = \frac{cos\ \delta}{sin\ a}\ sin\ t$$

The solar azimuth a is also counted positively with the course of the sun from the South=O°, through West toward North. When the radiation strikes a surface with a randomly oriented orientation, as in the practical application of this computation to topo-climatology, the circumstances become noticeably more challenging. If one considers the inclination of the surface with the angle v as being the rotation of a vertical wall by its horizontal central axis (v=90° would then correspond to the extreme case of the vertical wall, v = 0° to the opposite extreme case of the horizontal plane), then the wall azimuth 'á' of the vertical surface changes into the slope azimuth α (= 90°- á). The angle of incidence of the sloped surface, i.e., a slope, is then obtained through the general relation:

$$sin\ y = (sin\ \phi\ cos\ v - \ cos\ \phi\ \sin v\ cos\ \alpha) sin\ \delta + (cos\ \phi\ cos\ v + \ sin\ \varphi \sin v\ cos\ \alpha) cos\ \delta\ cos\ t + sin\ v\ sin\ \alpha\ cos\ \delta\ sin\ t$$

The extended Lambert formula is suitably used for the determination of the intensity of the sunshine I_m. Then:

$$I_m = I_0 \cdot e^{-a_m \cdot T \cdot m}$$

Where: I_o represents the solar constant. e = 2.718, the basis of the natural logarithms, a_m is the extinction coefficient (relative to the air amount m) T is the turbidity factor that was introduced by Linke and m is the amount of air trans-radiated. The strength of irradiation I on a randomly sloped surface that forms the angle y with the solar rays is then:

$$I = I_0 \cdot e^{-a_m \cdot T \cdot m} \cdot \sin y$$

The criteria for calculating the intensity of irradiation from direct sun radiation are now provided by this formula and the prior equation for sin-y. Given that such computations require a lot of work, one will now merely employ a computer. It is appropriate to employ short-period computer computations, such as 5-min steps, to determine the daily values of the strength of slope irradiation. These values should then be numerically integrated for the full day. This procedure only determines direct solar radiation under a cloudless sky, as can be observed from the formula provided for the computation of solar radiation intensity. There are still challenges in the calculation of topographically extremely different areas, for example, because the variances in the albedo are so tightly related to the values of the sky radiation. For topo-climatological purposes, however, determining the intensity of direct solar radiation should be sufficient because differences worth mentioning are only to be expected in the meteorological conditions in the lower air layers near the ground and in the heat conversion on the surfaces when direct solar radiation is present.

Naturally, the method for determining the intensity of irradiation provided here only produces the highest levels of solar radiation. First, the impact of clouds is not taken into account. Thus, despite potentially high solation values, a slope whose irradiation maximum occurs at a time of day when substantial clouding is typically present will only get weak radiation intensities. However, it can be difficult to analyze the impact of clouding because there aren't any good observations, especially for long periods of time. However, data on the length of sunshine can be used to draw an indirect inference about cloudy circumstances and the resulting reduction in radiation. A survey on the actual hourly energy application of a slope or of a horizontal surface that is sufficient for practical use can be obtained by multiplying the hourly sum of the calculated irradiation by the time-wise appropriate value of sunshine duration (in 1/10 h), assuming that the maximum radiative energy of a daytime hour can be recorded only during unlimited sunshine.

The Influence of Superelevated Surfaces

The superelevated horizon frequently plays a significant role in assessments of radiation conditions in orographically pronouncedly arranged terrane or in areas close to structures. Corresponding impediments shorten the duration of sunlight and, as a result, direct irradiation. It is simple to find the ecliptic's abbreviation. The strength of the irradiation to the appropriate surface must be subtracted from the energy decrease caused by the horizon's limitation. To illustrate the impact of horizon limitation for general direction, the table below gathers the dawn and sunset times for a North-South and for an East-West valley with varyingly high horizon superelevation. Horizon superelevation reduces global radiation by limiting the intensity of solar radiation as well as sky radiation. In addition to having an impact on how much energy is consumed during the day, horizon superelevation also plays a significant part in the radiation environment at night. There is an increase in the reflecting radiation and a corresponding decrease in the balance of long wavelength radiation, or of the effective radiation, in sloped surfaces that practically have their own horizon superelevation, as well as in dells and valleys.

Table 3: *Spectral distribution of global radiation*

	Wavelength range	**Strength of irradiation in cal cm- 2**	**Percent of the total strength of irradiation**
Ultraviolet		0.1	6.3
UV-B	0.28-0.315		0.35
UV-A	0.315-0.40		5.9
Visible light		0.8	52.1
	0.40-0.48		11.8
	0.48-0.60		18.2
	0.60-0.78		22.1
Infrared		0.6	41.6
	0.78-1.4		30.6
	1.4 -3.0		10.8

Source: Seemann et al., (1979).

Spectral Distribution of Global Radiation

It is well known that solar radiation is not uniform; rather, it is made up of radiation with different wavelengths. We divide this radiation into three categories: the ultraviolet (UV) portion, the visible radiation spectrum, which includes light, and the infrared spectrum (see Table above). One distinguishes between two ranges of ultraviolet radiation, each of which has distinct consequences. Human skin directly becomes pigmented by UV-A rays, while UV-B causes erythema. The agricultural meteorologist is especially concerned about the light. It makes up most of the solar radiation and, depending on the wavelength range, has quite diverse biological effects. The composition of light differs for the following color spectrums:

- Violet, 360-424 nm; blue, 424-492 nm; green, 492-535 nm; yellow, 535-586nm; orange, 586--647nm, and red, 647-760nm.

The ratio of direct to diffuse radiation affects how spectrally composed light is. When compared to direct radiation, diffuse radiation has a higher percentage of short-wavelength radiation (i.e., green and blue). Unless specific spectral bands are absent, the human eye perceives the spectrum of sunlight as "white". For instance, an item will appear to be green if a loss happens in the red spectral range, whereas it will appear to be red if a loss occurs in the green spectral region. The spectral sensitivity or effectiveness curve of the human eye peaks at about 555 nm. The strength of solar radiation, measured in lux (Ix) or lumen/m2 (Im/m2), was calculated from this effectiveness curve and the spectral distribution of solar radiation. According to the spectral sensitivity distribution of the human eye, the unit lux always refers to the entire spectrum range of light sensitivity. So, one may not use lux as a measuring unit when discussing the measurement of specific spectral ranges. The average light strength from extraterrestrial sources is 140,000 Ix. Naturally, clouds have a significant impact on both the intensity of illumination and global radiation. The average illumination under a cloudy sky will be only around 27% of that under a clear sky. However, It is feasible that reflection effects could lead to stronger illumination in the presence of dispersed clouds than in a clear sky.

The Biological Effects of Light

When everything is considered, the optical radiation in the earth's biosphere must be given significant weight. The following categories can be used to classify the specific biological effects:

Construction of organic substance

- photosynthesis
- formation of vitamin D
- formation of anthocyanins.

Transformation of matter

- pigmentation
- erythema formation
- bactericidal effects

Irritating effects

- photoperiodism
- phototropism
- phototaxis
- photo-nastic movement
- light- and darkness-germination
- photo-morphosis
- stimulation of nerves and glands.

Here, it is particularly important to discuss how light affects plant productivity. Naturally, photosynthesis occupies the foreground in this. In the chapter on greenhouse climates, photoperiodism—which is crucial to plant multiplication and the cultivation of decorative plants—will be explored in relation to artificial illumination. When one realizes that energy for photosynthesis in the order of around 8.6 1012 kcal per day is needed for the nutrition of people on Earth alone, one can see how important the energy output of solar light is to plant development. This enormous amount of energy is, in a manner of speaking, biochemically stored solar radiation. In photosynthesis, glucose is produced from carbon dioxide and water under bonding of energy from the light, whereby oxygen is being released:

$$6CO_2 + 6H_2O + 675KCal = C_6H_{12}O_6 + 6O_2$$

Sugar, carbohydrate, protein, and fat are then produced because of this primary assimilation. Only plants with organs that contain chloroplasts are capable of assimilating CO_2 (generally the leaves). Chlorophyll is responsible for the light's energy absorption. With rising light intensity, carbon dioxide assimilation rises under constant temperature and with an adequate air supply of CO_2. Although it does so more quickly at lower light intensities than at higher ones, this does not happen proportionally. There is no additional rise in photosynthesis after a certain level of light intensity. A portion of the surplus radiation is transformed into heat. The leaf is radiating with a reminder. The amount of “white” light that is transformed into chemical energy is just 0.6 7.7 percent. In typical shade-leaves as opposed to sun-leaves, the assimilation curve asymptotically decreases more quickly. In dim light, shade-leaves frequently assimilate a little bit more strongly than sun-leaves. The latter, however, are unquestionably better at stronger light intensities. The opposite process of respiration continues day and night in addition to the CO_2 - assimilation, which is known to occur exclusively in the presence of light. Respiration consumes matter whereas matter is generated via CO_2 absorption. When the amount of light falls below a particular point, respiration takes over during the day. The compensation point is the point at which the simultaneous production and consumption of assimilates are equal.

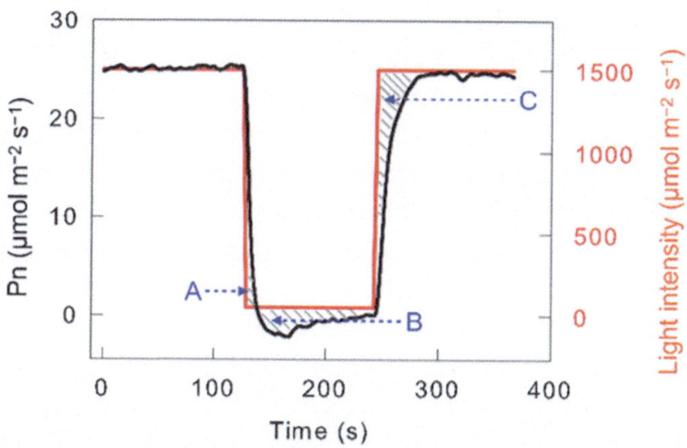

Figure 3: *Light intensity and dynamic net photosynthetic carbon assimilation rate (Pn) in leaves of C3 species Triticum aestivum. The A area indicates post-irradiance carbon assimilation; the B area indicates the post-irradiance carbon burst; the C area indicates the loss in photosynthetic carbon assimilation in the photosynthetic induction stage.*

Source: Li Y-T, Gao H-Y, Zhang Z-S. Effects of Environmental and Non-Environmental Factors on Dynamic Photosynthetic Carbon Assimilation in Leaves under Changing Light. Plants. 2023; 12(10):2015. https://doi.org/10.3390/plants12102015

The amount of carbon dioxide in the air entering and leaving the assimilation vessel, from which some carbon dioxide has been extracted during positive photosynthesis, serves as a restriction on its capacity. It is evident that the course of this curve is inverted during the morning and evening dark phases as well as at periods of insufficiently high light intensity. During these times, the plant emits carbon dioxide into the atmosphere as a result of its respiration. One can ascertain that the adjustment point has been exceeded by starting at a specific illumination level. The other case demonstrates that a "light saturation" had already happened at various periods, but the assimilation rate in the example of generally low light intensity responded very strongly to any change in light intensity. Beyond a certain point, the assimilation output essentially stops increasing even while the light intensity keeps increasing.

The uptake of CO_2 is energetically encouraged by specific light spectrum ranges. Therefore, the impact of light on the absorption of carbon dioxide depends not only on its intensity but also on its spectrum makeup. The assimilation peak is between 660 and 680 nm in the red. In the blue, between 470 and 550 nm, there is a secondary peak. The efficacy of wavelengths between 470 and 550 nm is quite low. This spectral distribution is in agreement with chlorophyll's capacity for absorption. Green- yellow is significantly less absorbent than red and blue. Additionally, the blue spectral region inhibits etiolation and promotes the synthesis of bio-growth components. Light is simply one factor that affects how much CO_2 is assimilated, as well as how quickly. The progression of this photochemical reaction is significantly influenced by temperature as well as the presence of carbon dioxide. Additionally, assimilation and respiration both occur at the same time and are solely temperature controlled. With rising temperatures, the respiration increases quite a little. Because CO_2 - assimilation is low due to the limited energy supply from light and respiration is too high as a result of high temperatures, photosynthesis, for instance, can go below the compensation point with just low light intensities and high temperatures.

Research illustrates how variable the light yield may be at differentially high temperatures. For example, at 10 °C 15,000 Ix produce an assimilation rate of about14 mm^3/h. The assimilation production increases to 35 mm^3/h at the same light intensity but at 30°C. Typically, carbon dioxide's impact won't be felt until absorption, when its concentration is drastically lowered. This is commonly the case when, despite other favorable circumstances, the carbon dioxide in the area around the plant is consumed and the resupply from the environment farther away does not occur quickly enough due to poor air circulation. Such a deficiency in carbon dioxide can occur in closed greenhouses as well as, under specific circumstances, in plants in the open countryside. In greenhouses, such a deficiency in carbon dioxide is frequently remedied by artificially raising the concentration (Seemann, 1965). Due to the rapid exchange of carbon dioxide molecules in an elevated partial pressure decrease in the atmosphere, such a technique can only be carried out with a certain amount of success in open terrain.

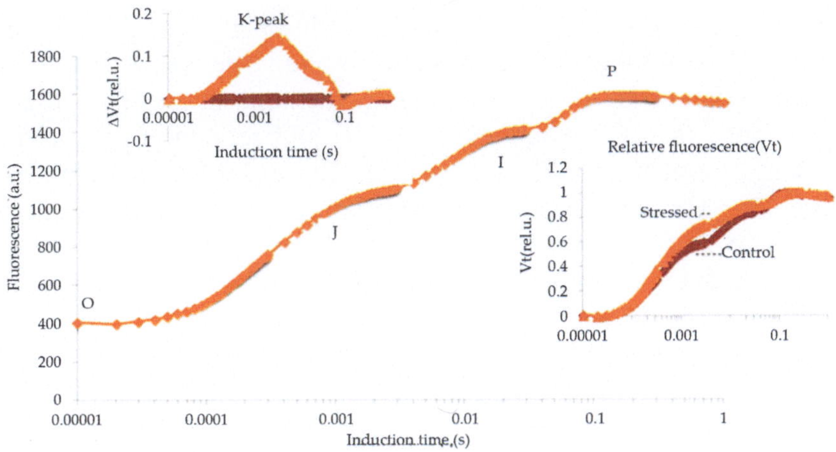

***Figure 4:** Chlorophyll* a *fluorescence parameters.*

Source: Janeeshma E, Johnson R, Amritha MS, Noble L, Aswathi KPR, Telesiński A, Kalaji HM, Auriga A, Puthur JT. Modulations in Chlorophyll a Fluorescence Based on Intensity and Spectral Variations of Light. International Journal of Molecular Sciences. 2022; 23(10):5599. https://doi.org/10.3390/ijms23105599

A rise in carbon dioxide may, under certain circumstances, result in absorption that is greater than usual. The cause of this is either that all other factors (light, temperature, etc.) are available for maximum assimilation in such a beneficial ratio that more carbon dioxide can be consumed than is typically available at normal exchange in the atmosphere, or that physiological conditions make the diffusion resistance for the entrance of carbon dioxide with normal partial pressure too great (closure or constriction of stomata or permeability limitations).

The fact that light as an energy source must be acknowledged to be the primary limiting factor in CO^{2-} assimilation under favorable temperature conditions and with a sufficient supply of carbon dioxide, however, cannot be changed in any of these considerations. Nothing else can take the place of light in this.

Net Radiation

For the radiation conversion of a surface, the absolute energy is always crucial. It can be transformed into another form of energy on its own, such as heat energy. It is well known that some of the solar energy that the sun emits and radiates toward the earth gets reflected, rendering it useless for converting it into other forms of energy. Additionally, energy is reflected from the surface of the ground or from plants and is then partially reradiated (reflected radiation) to the same surface by the atmosphere. The difference between long wavelength radiation (E + r) and reflection (A) is known as the long wavelength radiation balance. The difference between global radiation (S = H) and reflection (R), also known as albedo (net radiation). The ground's atmospheric reflection is reflected as r. Total radiative balance, or, more succinctly, net radiation, is the absolute radiative energy that can be transformed at the surface, for example, into heat (Q).

$$Q = (S + H) - R + A - E - r = (S + H - R) - (E - A + r)$$

Net radiation is, therefore = short wavelength less long wavelength balance. Basically, the short wavelength radiation balance is only functional when the sun is above the astronomical (or also orographic) horizon. The long wavelength radiation balance is effective the entire day. The short-wave balance is more susceptible to variations than the long-wave balance. The short wavelength balance significantly outperforms the long-wavelength balance in moderate latitudes during spring and summer, despite the long-wavelength balance producing bigger negative values during the day (since the surface temperature is higher during the day).

Therefore, under these circumstances, the net balance is positive during the daytime and negative at night. This is not to argue that there aren't any negative radiation balances before or after sunrise. Naturally, a number of variables affect the value of each radiation component. The surface conditions have the biggest impact on the reflection value. One distinguishes between diffuse and mirrored reflection in this regard.

The former occurs primarily on rough surfaces, while the mirrored reflection on smooth surfaces occurs on things like leaves with wax coatings and water. The value of reflection is somewhat influenced by the sun's angle. Additionally, wavelength affects reflection. Long wavelength radiation of individual formations is a function of the surface temperature (T_S). Stefan-Bolzmann's law:

$$E = \varepsilon\sigma T^4 \ Cal \cdot cm^{-2} \cdot min^{-1}$$

Depending on the surface, there may be slight variations (I: for grass, 0.97, for sand, 0.98, and for water, 0.95). Since the sum E + r (long wave radiation plus atmospheric reradiation reflection) is frequently measured in practice, when ε = 1, it equates to measurement accuracy. With the help of the table below, it is possible to calculate the intensity of rear body radiation (long wavelength radiative intensity) from the daily variations in surface temperatures.

***Table 4**: Long wavelength radiation into the hemisphere*

Temp. in °C	- 30	- 25	- 20	- 15	- 10	- 5	0
E+r	288	312	339	366	395	426	459
Temp. in°C	5	10	15	20	25	30	35
	493	530	568	609	651	696	743

Source: Seemann et al., (1979).

Schulze (1970) utilized the temperature from inside the weather observation hut (2 m) to determine E + r because it is typically impossible to monitor the surface temperature. He discovered the following connection:

$$E + r = 0.99\sigma T_{Air}^{4} in\ mcal \cdot cm^{-2} \cdot min^{-1}$$

This formula should be useful for order-of-magnitude estimates.

***Table 5:** Reflection on the surface in percent*

		Total	Light
Meadow	Dry	15-35	5-12
Grain	Green	15-25	7-15
Clover		17-32	
Fallow soil	Dry	8-15	
Sand	White, yellow	34-40	15-35
Loam	Dry	23	
Grey soil	Dry	25-30	11-20
Black	Dry	8-16	4-7
Needle-wood	Forest	10-14	3-7
Water surfaces		5-12	6-16

Source: Seemann et al., (1979).

The amount of water vapor in the atmosphere and the presence of clouds are the main determinants of atmospheric re-radiation. Its intensity is nearly never higher than the soil's long-wavelength radiation. For the purpose of calculating atmospheric counter radiation, Angstrom developed the experimentally verified formula shown below:

$$A = \sigma T_s^4 (\sigma = 0.82 - 0.25 + 10^{-0.2e})$$

In respect of the temperature in the weather observation center (T_{Air}) the reflection can be described in a simple manner as follows:

- for clear sky (daily average)

  $A = 0.82\sigma T_{Air}^4 \; cal \cdot cm^{-2} \cdot min^{-1}$

- for cloudy sky (daily average)

  $A = 0.94\sigma T_{Air}^4 \; cal \cdot cm^{-2} \cdot min^{-1}$

The daily course of reflected radiation does not have a particularly high daily average. The table below can roughly depict how cloud cover affects reflected radiation. It is founded on data gathered in Austria.

Table 6: *Daily totals of reflected radiation in 200 m above NN in the Austrian Alps in Cal·cm⁻²*

Month	*January*	*May*	*September*
Clouding0/10	450	627	631
5/10	505	678	682
10/10	600	771	776

Source: Seemann et al., (2012).

The aforementioned equation can be used to roughly compute the nighttime long wavelength net radiation. In those places:

$$-Q = (E - r) - A$$

for clear sky

$$= 0.162\sigma T_{Air}^4$$

for overcast sky

$$= 0.05\sigma T_{Air}^4$$

It is crucial to realize that when using these calculations, generic values for the resultingly estimated nightly radiation are all that may be expected.

The predicted re-radiation can vary from the measured one by 5%, and in rare circumstances by 10%, according to Bolz and Fritz (1950). Additionally, the temperatures of the air are being employed here, despite the fact that in terms of radiative laws, the temperature of the surface is what matters most about radiation. Even in the smallest space, the instant net radiation values typically fluctuate greatly, mostly due to variations in albedo and surface temperatures.

Radiation Measurement Technology

Agricultural meteorology generally requires the following measurements of solar radiation:

- Global and reflected radiation (albedo)
- Net radiation,
- Light
- Infrared radiation.

For measuring global radiation, pyranometers, or measuring devices with an angle of 1800, are nearly often utilized. Their sensitivity reaches approximately 3μ. The solarimeter, the Eppley pyranometer, the stellar pyranometer of Linke, and the actinograph of Robitzsch are the instruments that are used the most commonly, according to Moll-Gorczynski. The Moll-Gorczynski is a device with a black surface and an integrated thermos column that generates a thermal voltage proportionate to the intensity of the irradiation. Two glass culottes are concentrically positioned to cover the heat column. The zero points are only very little decreased in this way. The sensor surfaces of the black-and-white Eppley pyranometer are arranged in a circular arrangement. Black-and-white technology is also used in Linke's stellar pyranometer. It has black and white sensor surfaces arranged in a starry pattern. Thermos-columns can be used with Eppley- and stellar pyranometers as well. The Robitzsch bi-metal actinograph has proven itself in measurements when there aren't any very strict accuracy standards. This instrument is mechanical as opposed to the pyranometers discussed above. As radiation receivers, three bi-metal strips combine to create a flat surface. The core bimetal is black, while the outside two are white. These are counter-circuited to one another in order to account for temperature fluctuations brought on by factors other than radiation. Only the irradiation effect, which heats the black strip more intensely than the white strip due to higher absorption, causes the surface to curve. Lever methods are used to move it to the recorder. In this way, any variation in the radiation is recorded as a curve on a recording strip that is fastened to a drum. Also present above the measuring surface is a glass culottes. Registration using this instrument allows for a good overview of the radiation course in a place with typical maintenance. The actinograph's daily results are fairly consistent with one another. Along with zero-point stability, other important factors to consider are the stability of the standardization constant and, in particular, the best possible agreement with the "cosine law," which states that radiation incident at different angles of incidence should be as accurate as possible in its correspondence to the cosine of the angle of incidence. These are the factors that determine the pyranometer's quality. All wavelengths within the range of global radiation should be equally sensitive to the pyranometer at the same time. All of the other pyranometers mentioned here, with the exception of the actinograph, can be used for albedo measurements in addition to global radiation observations.

Only after the development of polyethylene (Lupolen H) for covering caps could net radiation meters be constructed in an efficient manner (Schulze, 1953). In a sense, the net radiation meter consists of two pyranometers, one of which is pointed upward and the other downward. The top measuring device monitors atmospheric reflection as well as solar and sky radiation (J+H). The downward measuring equipment receives the reflected surface reflection, the reflected atmosphere reflection, and the reflected global radiation (R). The polyethylene caps that are mounted over the pyranometers, in contrast to instruments with glass culottes, permit the measurement of the long wavelength range (0.3-60μ) in global radiation measurements. As a result, the net radiation meter is a device that can gauge all wavelength ranges associated with heat and solar radiation on earth.

Various types of net radiation meters are produced nowadays. The net radiation meter, according to Schulze, is the most well-known and technically priciest. The upper and lower thermoelectric piles are arranged in a differentiation circuit in practically all net radiation meters so that the value of the net radiation is displayed directly on the registration device or a related galvanometer. It is also feasible to measure short wavelength and long wavelength radiation individually using Schulze's net radiation meter. Only photoelectric instruments are useful for measuring light practically. One distinguishes between cells with exterior or internal photo-effects in this regard. The former category includes photocells, whereas the latter category includes photoresistors and photoelements. As photocells, vacuum cells with alkali cathodes are employed (Gorlich, 1951). The idea is that when metal surfaces are exposed to light, electrons are released. The electrons are removed toward the anode if one connects the positive pole of a voltage source to a second electrode (alkali) and the negative pole to the metal surface. a current flow brought on by exposure to light. For the measurement to be repeatable, there must be a relationship between photocurrent and radiation intensity. Regarding monochromatic light, vacuum cells can meet this condition as long as the illumination intensity is not excessive. For alkali cathodes, cadmium, potassium, rubidium, and cesium are typically employed. Cadmium cells are used for UV measurements. The spectral sensitivity of potassium reaches from about 0.3 to 0.67 μ with a peak at 0.4μ. Cesium covers the wavelength ranges between 0.3 and 0.73 μ with a minimum at 0.4 and a peak at 0.56~. The sensitivity of rubidium ranges between 0.4 and 0.6 μ with a peak at 0.48μ.

Photocells can experience brief alterations as a result of age and exhaustion. Unlike fatigue, which is a transitory alteration brought on by heavy or prolonged loads that goes away when the cell is not in use, aging is an irreversible, permanent change. Photoelements have a wide range of applications, whereas photoresistors are scarcely usable for agriculture-meteorological applications due to their extremely limited outside applicability. The photoelement is made up of a thin, light-permeable front electrode and a metallic base plate with a layer of silicon or selenium placed over it (Lange, 1939, 1940). The front electrode, from which the current is drawn using a sprayed-on metallic ring, symbolizes the negative cell-pole while the base plate serves as the positive cell-pole. Light strikes the cell, producing electrons in the silicon or selenium, which subsequently flow out through the front electrode. In direct proportion to the brightness of the light, the photocurrent increases. One can get roughly 10 times more current using silicon than one can with selenium. The range of the selenium cell's spectral sensitivity is 350 to 750 nm, with a peak at 550 nm. It basically matches the sensitivity of the human eye in this regard. The peak of the sensitivity in the silicon cell is between 800 and 900 nm, and it extends farther into the long wavelength range (1170 nm). Dark resistance is comparatively high in silicon cells (about 500 kOhms). Because of this, they are not as effective as selenium cells when measuring agricultural meteorological light.

Manufacturers typically refer to equipment using selenium cells as "luxmeters" and standardize the measured values in lux for light measurements. After some time, the electrical stability of selenium photocells exhibits a decrease in sensitivity. Worner (1955) and others have conducted investigations into this. Therefore, more frequent standardization is advised. The simultaneous monitoring of solar radiation with a pyranometer offers some control over this. An exact conversion from lux to absolute units (Cal) is possible with the knowledge of the color temperature of the radiator. In the case of a temperature of about 5800 K (sun), 1lux corresponds to 1lx = 1.146 × 10^{-5} cal·cm^{-2}·min^{-1}. As was already established, the selenium photocell's spectral sensitivity is similar to that of the human eye, making it an excellent choice as a tool for measuring the intensity of illumination. Spectral sensitivity presents certain challenges for agricultural meteorological applications. It is difficult to compare the light output of artificial light sources with different spectral compositions. When taking into account the fact that the photosynthetically active radiation within the spectrum has a completely different intensity distribution than that detected by the selenium cell, it is also practical to only regard the measurement values in outdoor measurements as relative values. Because of this, unique measurement techniques have been created for photosynthetic applications. A filter combination (Schott-filters) created by Schulze in 1949 enables measurements with selenium cells in the chlorophyll absorption spectrum.

The measurement tool created by Lambda Instruments Corp. (Lincoln, Nebraska, USA) measures the wavelength range between 380 and 740 nm far more effectively than the selenium cell does. This sensor is particularly well suited for light-dependent research related to photosynthesis. Direct measurement of photosynthetically active radiation is achievable with this equipment. Micro-Einsteins (1 micro-Einstein = 6.023.1017 photons) are used as the unit of measurement. Briefly stated, the fundamental idea behind this measurement system is as follows: First, one must take into account the fact that only absorbed radiation currents can become energetically effective. Reference is made more challenging by reflection and transmission caused by different leaf positions and sun angles. Second, photosynthesis is an example of a physicochemical process or the conversion of physical energy into biological energy. Thirdly, electromagnetic radiation can be viewed as corpuscular, or as being made up of discrete quanta. Einstein's Law on photochemical equivalence would give the mental pre-requisite for the solution of the energy conversion problem in this scenario. According to this, there exists a simple integral relationship between the number of the molecules that are photochemically changed (chemical energy acceptance) and the number of the absorbed quanta in the range between 0.4 and 0.7 μ. Currently, the latter are referred to as photons. This makes it clear that a photon's energy level is irrelevant to photosynthesis, meaning that all photons within the aforementioned spectrum range contribute equally to the creation of matter.

It is known that the relative energy content of a photon in the range of 0.7 μ is only about 4/7 of the energy amount of a photon in the 0.4μ range. However, photons of shorter wavelength ranges contribute more significantly to heat dissipation than do photons of higher wavelength ranges because energy amounts that are not necessary for photosynthesis are transformed into heat. This is because electrochemical molecular changes only require constant amounts of energy. It is all that is required to count the number of photons in the visible wavelength region for the correlation of photosynthetic materials. The photon stream in the range between 0.4 and 0.7 μ is counted as the photosynthetically active solar radiation that is being used exclusively for photosynthesis. Energy that is not utilized photosynthetically is dissipated as heat. This definition is, among others, in agreement with McCree (1972, 1973) and Bjorkmann (1968), who also found a quasi-constant yield in the spectral range between 0.4 and 0.7 μ in using photon numbers. Gabrielson (1960) had made a suggestion along these lines. On this definition, there is still no universal agreement.

One should not ignore the fact that infrared radiation measurements have long been gaining importance when talking about radiation-measuring equipment for agricultural meteorological reasons. These are virtually primarily employed in agricultural meteorology for contact-free surface temperature determination.

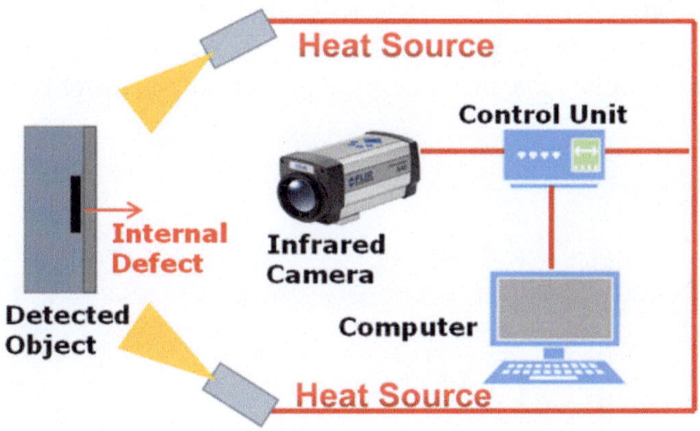

Figure 5: *Schematic diagram of an infrared thermography camera.*

Source: Qu Z, Jiang P, Zhang W. Development and Application of Infrared Thermography Non-Destructive Testing Techniques. Sensors. 2020; 20(14):3851. https://doi.org/10.3390/s20143851

A block schematic (see image above) can help in visualizing how such an infrared radiation measuring device is built and how it works. An infrared transparent lens and an edge filter are used to transmit radiation from a surface (a measuring object) to a radiation receiver (bolometer). A vibrating modulator that is fixed in front of the bolometer's diaphragm aperture intermittently interrupts the object radiation at a frequency of 35 cycles. The modulator's radiation, which is dependent on the temperature of the modulator, is detected by the bolometer during each interruption. The difference between the two radiation fluxes as a function of the object- and modulator-temperature determines the value of the output voltage, which is an alternating voltage. The output voltage is then made independent of the modulator temperature, i.e., dependent only on the object temperature, by adding a compensation step after amplification, phase-dependent rectification, and smoothing. The sensitivity is tuned for the wavelength range above 8 mil in order to effectively exclude the additional solar radiation reflected off the object. As a result, such a device is particularly well suited for contactless temperature measurements outside. Temperature resolution is at approximately ±0.25 °C in the range between - 20 and +60°C.

Pyranometer

Typical pyranometers don't need power to function. The use of electronics in pyranometers, which do require (minimal) external power, is a recent breakthrough in technology. The wavelength range of the solar radiation spectrum that reaches the surface of the planet ranges roughly from 300 nm to 2800 nm.

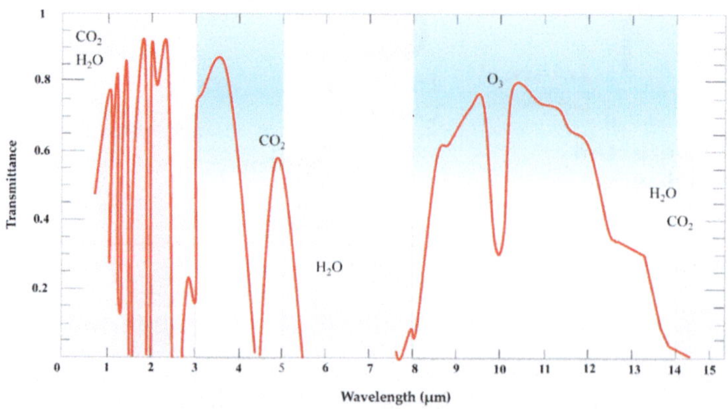

Figure 6: Atmospheric transmittance in the thermal region with typical absorption bands induced by gasses and water.

Irradiance readings will be produced with varying levels of spectral sensitivity depending on the type of pyranometer employed. By definition, the reaction to "beam" radiation must vary with the cosine of the angle of incidence in order to be measured for irradiance. This guarantees a full reaction when the sun shines directly on the sensor (normal to the surface, sun at zenith, 0° angle of incidence), a zero response when the sun is at the horizon (90° angle of incidence, 90° zenith angle), and a response of 0.5 when the sun is at a 60° angle of incidence. As a result, a pyranometer's "directional response" or "cosine response" should be as close to the ideal cosine characteristic as is feasible. Three types of pyranometers can be identified and categorized into two distinct technologies, thermopile technology and silicon semiconductor technology, using the definitions listed in the ISO 9060. The type of pyranometer determines the 'spectral response,' or light sensitivity.

Figure 7: *Pyranometer.*

Source: By Arthurcala - Own work, CC BY-SA 3.0, https://commons.wikimedia.org/w/index.php?curid=106656274

Thermopile Pyranometers

A sensor based on thermopiles known as a thermopile pyranometer, also known as a thermo-electric pyranometer, is used to measure the broad band of the solar radiation flux density from a 180° field of view angle. Thus, a thermopile pyranometer typically measures with a mostly flat spectral sensitivity between 300 and 2800 nm. The active portion of the sensor was evenly divided into black and white sectors in the first generation of thermopile pyranometers. The difference in temperature between the white sections, which are better described as being in the shade or not exposed to the sun, and the black sectors, which are exposed to the sun, was used to compute irradiation. Irradiation is proportionate to the difference between the temperature of the sun-exposed area and the temperature of the shade area in all thermopile technologies. A thermopile pyranometer is made of the following key elements in order to achieve the correct directional and spectral characteristics:

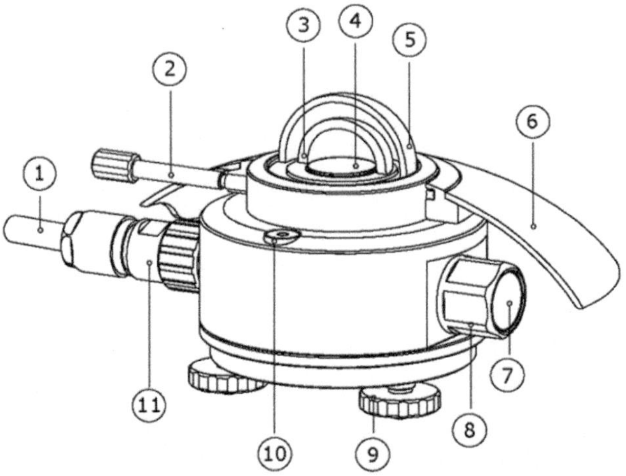

***Figure 8**: Schematic depiction of a pyranometer: (1) cable, (3) pyranometer and (5) glass domes, (4) black detector surface, (6) sun screen, (7) desiccant indicator, (9) levelling feet, (10) bubble level, (11) connector.*

Source: https://en.wikipedia.org/wiki/File:SR20_pyranometer_linedrawing.pdf

- **A thermopile sensor with a black coating:** It possesses an almost perfect cosine response, a flat spectrum encompassing the range of 300 to 50,000 nanometers, and it absorbs all solar light.
- **A glass dome:** It keeps the 180° field of view while restricting the spectral response to 300 to 2,800 nanometers (leaving off the portion over 2,800 nm). The thermopile sensor is additionally protected from convection by it. A second glass dome acts as an

additional "radiation shield" in many, but not all, first-class and secondary standard pyranometers (see ISO 9060 classification of thermopile pyranometers), which improves the thermal equilibrium between the sensor and inner dome in comparison to some single dome models by the same manufacturer. In these scenarios, having a second dome has the effect of significantly reducing instrument offsets. Class A, single dome models, with low zero-offset (+/- 1 W/m^2) are available.

The active (hot) connections of the thermopile in contemporary thermopile pyranometers are situated underneath the black coating surface and are heated by the radiation absorbed from the black coating. The pyranometer housing, which acts as a heat-sink, is in thermal contact with the passive (cold) connections of the thermopile, which are completely shielded from solar radiation. This avoids any alteration from yellowing or decay while taking a temperature reading in the shadow, which would affect how much sunlight is being received. According to the temperature difference between the black coated surface and the instrument housing, the thermopile produces a little voltage. This is of the order of 10 μV (microvolts) per W/m2, so on a sunny day the output will be around 10 mV (millivolts). Unless otherwise equipped with electronics for signal calibration, each pyranometer has a distinct sensitivity. Thermopile pyranometers are frequently employed in the fields of meteorology, climatology, research on climate change, building engineering physics, photovoltaic systems, and photovoltaic power plant monitoring. In meteorological stations, they are normally mounted horizontally. The solar energy sector has established the types of pyranometers that should be used based on the size and classification of solar power plants in a 2017 standard called IEC 61724-1:2017. To improve accuracy in performance ratio computation, that standard advises installing thermopile pyranometers horizontally (GHI, Global Horizontal Irradiation) and photovoltaic pyranometers on the plane of PV modules (POA, Plane-Of Array).

Photovoltaic Pyranometer – Silicon Photodiode

A photodiode-based pyranometer, also referred to as a photoelectric pyranometer in the ISO 9060, can detect the region of the solar spectrum between 400 nm and 1100 nm. The photoelectric effect enables the photodiode to rapidly convert the aforementioned solar spectrum frequencies into electricity. The temperature has an impact on the conversion since a rise in temperature results in an increase in current of approximately 0.1 percent per °C. A housing dome, a photodiode, and a diffuser or optical filters make up a photodiode-based pyranometer. The photodiode functions as a sensor and has a modest surface area. The photodiode produces a current that is proportional to the amount of light it receives; an output circuit, like a transimpedance amplifier, produces a voltage that is also directly proportional to the photocurrent. The output is typically in the millivolt range, which is the same range as thermopile-type pyranometers. When calculating the amount of radiation from the visible solar spectrum or from specific regions like the UV, IR, or PAR (photosynthetically active radiation), photodiode-based pyranometers are used. Utilizing diodes with certain spectrum sensitivities allows for this. The main component of luxmeters used in photography, film, and lighting method are photodiode-based pyranometers. Additionally, they are occasionally put next to photovoltaic system modules.

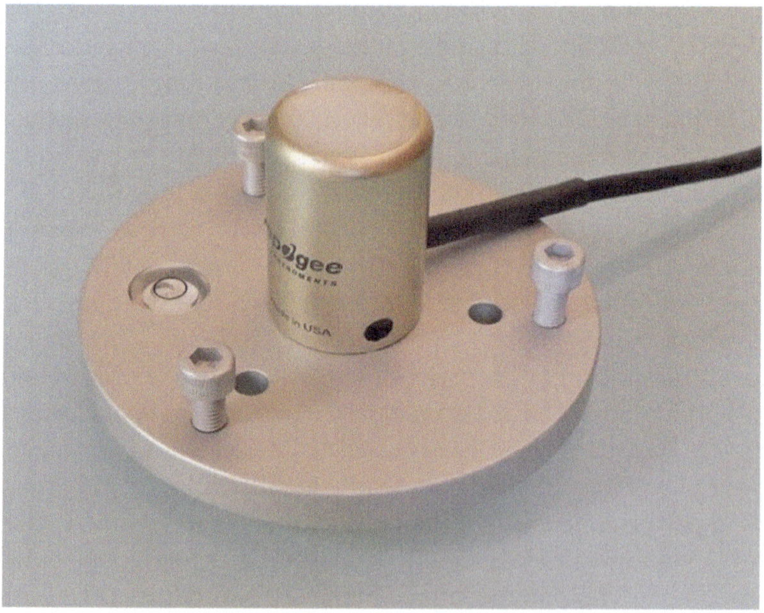

Figure 9: *Photodiode Pyranometer.*

Source: https://commons.wikimedia.org/wiki/File:Photodiode_pyranometer.jpg

Photovoltaic Pyranometer – Photovoltaic Cell

The photovoltaic pyranometer, which was developed in the 2000s alongside the proliferation of solar systems, is a development of the photodiode pyranometer. It met the requirement for a single reference photovoltaic cell when gauging the output of solar cells and modules. The producers of each cell and module specifically test them using flash tests, and thermopile pyranometers lack the necessary response time and spectral response of a cell. When measuring power, which would need to be quantified, this would result in an obvious disparity. This pyranometer's other name in the technical documentation is "reference cell." A solar cell that is operating in a close to short-circuit situation makes up the sensor's active component. As a result, the generated current is directly correlated with the solar energy that strikes the cell between the wavelengths of 350 nm and 1150 nm. It generates current as a result of the photovoltaic effect when exposed to light in the above range. Although it is not flat, its sensitivity is comparable to that of a silicon photovoltaic cell. The following components are used to assemble a photovoltaic pyranometer:

- A metallic container with a fixing staff
- A small photovoltaic cell
- Signal conditioning electronics

Silicon sensors with temperature-dependent outputs include photodiodes and photovoltaic cells. The electronics in more current models adjust the signal according to temperature, eliminating the temperature effect from solar irradiance values.

The casing of some models houses a board for the signal's amplification and conditioning. For calculating photovoltaic module effective power and system performance, photovoltaic pyranometers are used in solar simulators and alongside photovoltaic systems.

A photovoltaic pyranometer can also be used to make a first diagnosis of a failure in solar systems since its spectrum response resembles that of a photovoltaic module.

In order to link the Module Temperature Sensor, Ambient Temperature Sensor, and Wind Speed Sensor with just one Modbus RTU output directly connected to the Datalogger, the Reference PV Cell or Solar Irradiance Sensor may have external inputs. The Solar PV Plants may be monitored using these data.

Pyranometer Standardization and Calibration

Both thermopile-type and photovoltaic pyranometers are manufactured according to standards.

Thermopile Pyranometers

The World Meteorological Organization and thermopile pyranometers both adhere to the ISO 9060 standard (WMO). Three classes are separated by this standard.

As of 2018, the most recent version of ISO 9060 utilizes the following classification: While the former ISO 9060 standard from 1990 utilized misleading words like "secondary standard," "first class," and "second class," the newer ISO 9060 standard uses Class A for the best performing, followed by Class-B and Class C. Response time, thermal offsets, temperature dependence, directional inaccuracy, non-stability, non-linearity, spectral selectivity, and tilt response are a few of the sensor's characteristics that contribute to class differences. Each of them is described in ISO 9060. A sensor must satisfy all necessary conditions for these features in order to fall under a specific category. Spectral flatness and fast responsiveness are two sub-categories that are part of ISO 9060:2018. They aid in better differentiating and classifying sensors. To gain the 'fast response' classification, the response time for 95% of readings must be less than 0.5 seconds; while 'spectrally flat' can apply to sensors with a spectral selectivity of less than 3% in the 0.35 to 1.5 μm spectral range. While the majority of Class-A pyranometers are "spectrally flat," fast response sensors are far less common. Response times for the majority of Class-A pyranometers are 5 seconds or longer. The World Radiometric Reference (WRR) is frequently used as an absolute reference during calibration. In Davos, Switzerland, PMOD looks after it. Other than the World Radiometric Reference, private labs with accreditation for these particular calibrations include ISO-Cal North America. Calibration for the Class A pyranometer is carried out in accordance with ASTM G167, ISO 9847, or ISO 9846. Typically, class-B and class-C pyranometers are calibrated in accordance with ISO 9847 and ASTM E824.

Photovoltaic Pyranometer

In accordance with IEC 60904-4 for primary reference samples and IEC 60904-2 for secondary reference samples and instruments that are meant for sale, photovoltaic pyranometers are standardized and calibrated. Both standards have a primary standard known as the group of cavity radiometer by the World Radiometric Reference (WRR) that serves as the beginning of their respective traceability chains.

Pyranometer Signal Conditioning

Typically, the natural output values of these pyranometers don't surpass tens of millivolts (mV). Since it is a "weak" signal, it is particularly susceptible to electromagnetic interferences when the cable travels over large distances or is installed near photovoltaic installations. As a result, these sensors typically have signal conditioning electronics installed, which provide an output of 4–20 mA or 0–1 V. An alternative approach suggests improved noise immunity, such as Modbus via RS-485, appropriate for environments with electromagnetic interferences typical of medium-large solar power stations, or SDI-12 output, where sensors are part of a low power weather station. The equipped electronics frequently agree to simple SCADA integration. The electronics of the sensor can additionally store additional data, such as the serial number and calibration history.

CHAPTER 3

SURFACE ENERGY BALANCE

CONTENTS

OVERVIEW

The use of a model makes the processes involved in heat conversion on the soil the most understandable. The positive net radiation that is transformed into heat on the surface is led away in many ways if one considers a portion of the earth's surface that is devoid of vegetation and may be generally considered to be flat. The ground warms up as a result of some of the heat being led into the ground. A further fraction is used for air heating. The sensible heat stream is the name given to this part. When there is water at the surface, a third of the heat conversion is used for vaporization. The vapor has this heat latently retained. Heat cannot be stored since the energy conversion from net radiation occurs on a surface, therefore the total energy is distributed across the three heat streams. This implies that net radiation (Q) less heat flux into the soil (B), less heat transfer to the air (L), and less heat consumed in vaporization (V), expressed in a straightforward equation, must equal zero;

$$Q - B - L - V = O$$

The heat home equation is the name of this energy equation. An additional heat flux is added to the "plant bodies" if the presumed earth's surface is vegetated. The process of photosynthesis also uses up additional energy. However, this fraction is so negligible in contrast to the other energy fluxes that it can be disregarded.

The heat fluxes are inverted in the event of a negative net radiation. Heat is released to the surface by the soil and the air. Additionally, condensation allows the latent heat contained in the water vapor to escape to the surface (formation of dew). The heat that has been released to the surface is subsequently transformed once more into radiation energy. Therefore, the domestic heat equation at night is;

$$-Q + B + L + V = O$$

The heat fluxes are, in other words, directed towards the surface in the case of negative net radiation whereas they are directed in the opposite direction in the case of positive net radiation. That is the standard. The soil heat flux can occasionally, nonetheless, show a positive sign throughout the day and drive heat to the surface. The heat household equation and its explanation of the model concept for energy conversion has particular practical significance for agricultural meteorology. It enables the skilled agricultural meteorologist to estimate the magnitudes of temperature conditions and water conversion in the widest range of circumstances. However, the determination of the actual evapotranspiration, which will be covered later, places a specific emphasis on this model.

CONTACT POINT

The line separating the atmosphere from the land or water is the Earth's surface. It can be challenging to pinpoint the location of this border over an extremely disturbed sea or over land surfaces with a changeable plant canopy. While treating the surface as a straightforward interface between two media and assuming that its location can be accurately determined, we must also take into account the atmosphere, oceanic boundary layers, and the top few meters of soil when thinking about the significant energy exchange processes. Given that the climate at the surface has the greatest practical impact, the energy fluxes over the surface are just as significant to the climate as the fluxes at the top of the atmosphere. The amount of energy flux available to evaporate surface water and raise or drop the surface temperature is determined by the surface energy balance. Surface processes are also crucial in defining the planet's overall energy balance. Compared to the budget at the top of the atmosphere, the energy budget near the surface is more complicated because it also takes into account fluxes of energy through conduction, convection of heat and moisture through fluid motion, and radiation. The local surface energy budget is influenced by the insolation, surface properties including wetness, vegetation cover, and albedo, as well as the properties of the surrounding atmosphere. Since evaporation from the surface plays a significant role in both the energy and water budgets, the hydrologic cycle and the surface energy budget are closely related. Understanding climate and its reliance on external restrictions requires an understanding of the energy budget of the surface.

THE SURFACE ENERGY BUDGET

The energy budget is measured in watts per square meter and can be expressed in terms of energy flux per unit area moving vertically across the air-surface interface. Solar and infrared radiative transfer, as well as energy fluxes linked to fluid dynamics in the atmosphere and oceans, are the mechanisms that control energy transmission between the surface and the atmosphere. Energy transportation and storage below the surface are also significant. For the purposes of calculating energy budgets, surface storage occurs in the area between the atmosphere's boundary and a depth below the surface where energy fluxes and the rate at which energy is stored are deemed insignificant. This depth can range from a few meters on dry land to several kilometers where deep water is generated in the ocean. The horizontal energy fluxes produced by the fluid motions under the surface can be crucial for water surfaces. It is possible to write the surface energy balance symbolically as;

$$\frac{\partial E_s}{\partial t} = G = R_s - LE - SH - \Delta F_{eo}$$

where $\partial E_s/\partial t = G$ is the storage of energy in the surface soil and water, R_s is the net radiative flux of energy into the surface, LE is the latent heat flux from the surface to the atmosphere, SH is the sensible heat flux from the surface to the atmosphere, and ΔF_{eo} is the horizontal flux out of the column of land ocean below the surface. Under steady-state conditions in which the storage of energy is small, such as one might assume to hold for annual averages or for daily averages over land, the energy balance is between radiative heating and the processes that remove energy from the surface.

$$R_s = LE + SH + \Delta F_{eo}$$

The radiative, latent, and sensible heat flux terms are typically positive because, under most circumstances, radiation heats the surface while latent and sensible heat fluxes cool it. The energy storage beneath the surface is equal to the net radiative input less the heat lost by evaporation, sensible heat flux, and horizontal heat transport to different latitudes or longitudes. This is the physical meaning of the graph below. Numerous more terms that may be significant locally or for short periods of time were used in the construction of the picture below. These include the following terms:

- It is possible that 10% of the radiative imbalance will be needed for a brief period of time to generate the latent heat of fusion needed to melt springtime ice and snow.
- In general, only a small amount of wind and wave kinetic energy is converted to thermal energy.
- If the precipitation is warmer than the surface, heat transfer through precipitation may take place. Due to the huge thermal capacity of water and the potential for significantly cooler precipitation than the surface during summer showers, this mechanism is especially crucial.
- Some of the solar energy is kept in the chemical bonds created during photosynthesis rather than being converted to heat. This is less than 1% on a global scale, but can occasionally fall below -5% locally.
- The opposite of photosynthesis is the release of heat caused by the oxidation of biological materials, such as in biological

decomposition or forest fires. Through oxidation, the energy stored in biological matter during photosynthesis is released back into the physical climate system. Although it happens all year long, this process moves forward more quickly when the surface is warm and damp.

- Hot springs, earthquakes, and volcanoes all only emit a little amount of geothermal energy overall.
- Although the heat produced by burning fossil fuels or producing nuclear power may be enormous locally, it has little impact on the world's energy balance.

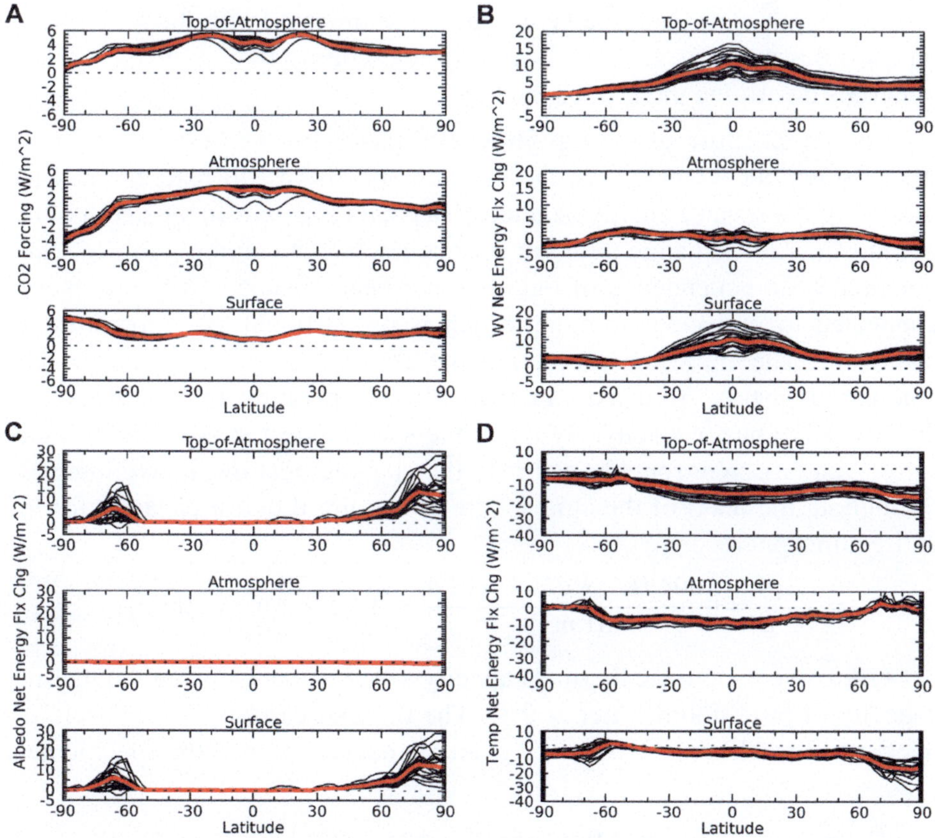

Figure 10: *Diagram showing the relationship of the various terms in the surface energy balance.*

Source: Sejas SA, Hu X, Cai M and Fan H (2021) Understanding the Differences Between TOA and Surface Energy Budget Attributions of Surface Warming. Front. Earth Sci. 9:725816. doi: 10.3389/feart.2021.725816

Storage of Heat in the Surface

For the seasonal cycle of temperature over the oceans and the diurnal cycle over land and ocean, energy storage in the surface is crucial. The amount of energy at the surface can be calculated by multiplying the effective heat capacity of the Earth-ocean system by the corresponding mean temperature.

$$E_s = \hat{C}_{eo} T_{eo}$$

Where $\hat{C}_{eo}$ is the effective heat capacity of the land or ocean system ($Jm^{-2}K^{-1}$) and T_{eo} is the effective temperature of the land or ocean energy-storing material (K).

The depth of the surface layer that communicates with the atmosphere on the relevant time scale and the physical characteristics of the surface materials determine the heat capacity.

The temperature of the top 50 to 100 meters of the ocean varies with the seasons, but typically just the first few meters of soil react to seasonal forcing of the surface energy balance. Seasonal variations in the depth of the ocean's energy exchange with the surface necessitate taking into account a potential time dependence of the effective heat capacity. Only the energy connected to molecular motion, which is proportional to temperature, is taken into account when estimating the atmosphere's heat capacity. The amount of energy needed to increase the temperature by one degree is known as the heat capacity. By using the specific heat at constant pressure for dry air, we may roughly estimate the specific heat of air. We integrate throughout the mass of the atmosphere to acquire the heat capacity for the entire atmosphere.

$$\hat{C}_a = C_p \frac{p_s}{g} = \frac{1004\, JK^{-1}Kg^{-1} \cdot 10^5 Pa}{9.81\, ms^{-2}} = 1.02 \times 10^7 JK^{-1}m^{-2}$$

One can estimate the thermal capacity of the ocean by using the thermal capacity of pure liquid water at 0°C. The thermal capacity for an arbitrary depth of water d_w, can be obtained from the density ρ_w, and the specific heat c_w.

$$\hat{C}_o = \rho_w\, c_w\, d_w = 10^3 kg\, m^{-3} \cdot 4218\, JK^{-1}Kg^{-1} \cdot d_w$$

$$= d_w \cdot 4.2 \times 10^6 kg\, m^{-3} \cdot 4218\, JK^{-1}\, m^{-2} m^{-1} \cdot d_w$$

Comparing and we see that the thermal capacity of the atmosphere is equal to that of a little over 2 m of water.

Heat Storage in Soil

Compared to the ocean, the land has a substantially lower effective heat capacity. Because the surface is solid, there is no efficient heat transfer through fluid motions like there is in the ocean and atmosphere. The less effective conduction pathway is primarily responsible for the transport of heat through the soil. Seasonal changes only impact the top meter or two of the soil. A ground surface normally has a slightly lower heat capacity than the atmosphere. The relationship between the vertical temperature gradient in the soil and the vertical flux of energy by conduction in the soil is linear.

$$F_s = -K_T \frac{\partial T}{\partial z}$$

The thermal conductivity is denoted by K_T. Storage of heat in the soil and convergence of the diffusive heat flux make up the soil's thermal balance.

$$C_s \frac{\partial T}{\partial t} = -\frac{\partial}{\partial z}(F_s) = \frac{\partial}{\partial z}\left(K_T \frac{\partial T}{\partial z}\right)$$

The volumetric heat capacity of the surface material C_s is the product of the specific heat of the soil c_s, and the soil density ρ_s. The heat capacity of the soil depends on the volume fractions of soil f_s, organicmatterf_c, water f_w and air fa, and the density and specific heat of each component of the surface material.

$$C_s = \rho_s c_s f_s + \rho_c c_c f_c + \rho_w c_w f_w + \rho c_p f_a$$

***Table** 7: Properties of Soil Components at 293 K*

	Specific heat (C_p) JK^{-1}K^{g}-1	**Density (ρ) (Kg m^{-3})**	ρC_p
Soil inorganic material	733	2600	1.9 x 10^6
Soil organic material	1921	1300	2.5 x 10^6
Water	4182	1000	4.2 x 10^6
Air	1004	1.2	1.2 x 10^3

Source: Brutsaert (1982).

Since the soil air has a very low heat capacity, as can be seen from the table above, the heat capacity of soil increases significantly when water is introduced to fill the open spaces. The volumetric portion of the soil that can be occupied by air or water is known as porosity. Material, porosity, and soil water content all affect a soil's ability to transmit heat. For soils with reasonably high porosity, the thermal conductivity rises with the amount of water present. Values vary from 0.1 $Wm^{-1}K^{-1}$ for dry peat to 2.5 $Wm^{-1}K^{-1}$ for wet sand. Under the condition that the thermal conductivity, K_T, is independent of depth, simplifies to the heat equation;

$$\frac{\partial T}{\partial t} = D_T \frac{\partial^2 T}{\partial z^2}$$

where $D_T = K_T/C_s$ is the thermal diffusivity of the surface material. A simple scale analysis can be used to determine the depth through which a temperature anomaly applied at the surface will penetrate in a given time. One can show that the penetration depth, h_T, of temperature anomalies associated with a periodic forcing of temperature at the surface is given by;

$$h_T = \sqrt{D_T \tau}$$

where z is the time scale of the periodic forcing at the surface. Taking a typical value of soil diffusivity of $D_T = 5 \times 10^{-7} m^2 s^{-1}$, we obtain a penetration depth of about 10cm for diurnal forcing and about 1.5m for annual forcing. The surface material would be penetrated to a depth of around 150 m by a change in surface temperature over 10,000 years. Horizontal heat movement beneath the land surface is completely negligible due to the poor rates of heat conduction through rock and soil. The graphs below depict the temperature distribution in the soil as a function of time on a clear summer day.

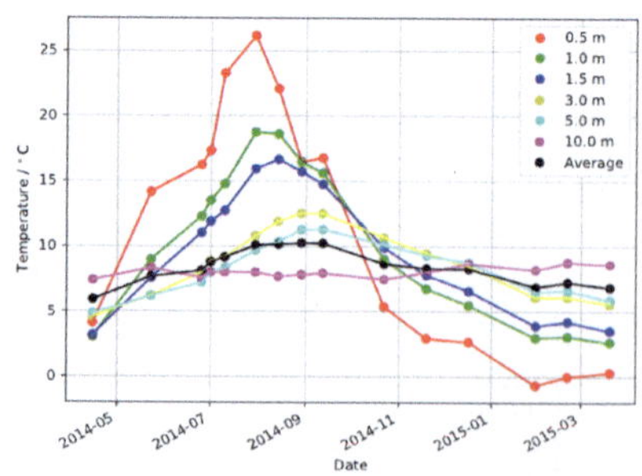

***Figure 11:** Soil temperatures measured at different depths.*

Source: Çuhac C, Mäkiranta A, Välisuo P, Hiltunen E, Elmusrati M. Temperature Measurements on a Solar and Low Enthalpy Geothermal Open-Air Asphalt Surface Platform in a Cold Climate Region. Energies. 2020; 13(4):979. https://doi.org/10.3390/en13040979

The temperature of the soil close to the surface varies greatly during the day, reaching its highest point around after midday and its lowest point immediately before daybreak. Due to the time it takes for the temperature pulse to penetrate into the soil, temperature changes are smaller and occur later in the day deeper in the soil. By around 10 cm below the surface, the amplitude of the temperature perturbation is roughly e-1 of its surface value, and by 40 cm below the surface, it is very modest. The above-mentioned straightforward dimensional arguments are supported by the observed decrease in the amplitude of the daily temperature perturbation.

The temperature data can be used to infer the thermal conductivity as a function of depth in the soil. Alternatively, if the vertical profiles of C_s and K_T are known and assumed constant with time, then measured temperature profiles in a deep layer of Earth can be used to estimate past variations in surface temperature on time scales of hundreds to thousands of years.

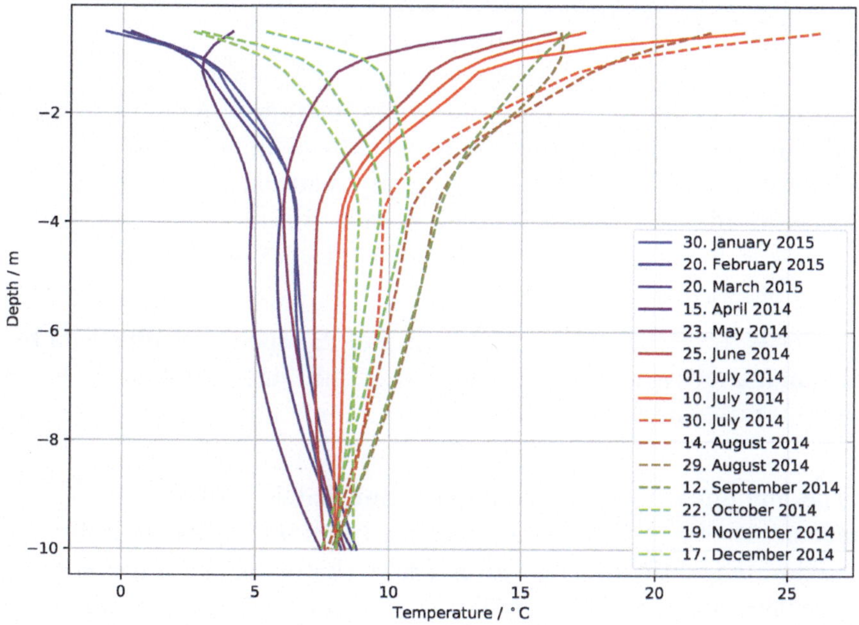

***Figure 12:** Seasonal soil temperatures at different depths.*

Source: Çuhac C, Mäkiranta A, Välisuo P, Hiltunen E, Elmusrati M. Temperature Measurements on a Solar and Low Enthalpy Geothermal Open-Air Asphalt Surface Platform in a Cold Climate Region. Energies. 2020; 13(4):979. https://doi.org/10.3390/en13040979

Radiative Heating of the Surface

The net input of radiative energy to the surface is the sum of the net solar and longwave flux densities at the surface.

$$R_s = S\downarrow(0) - S\uparrow(0) + F\downarrow(0) - F\uparrow(0)$$

where S↓ (0) and S↑ (0) are the downward and upward flux density of solar radiation at the surface, respectively, F↓ (0)and F↑ (0) represent similarly defined longwave flux densities. Solar heating is the basic driver of the climate system.

Absorption of Solar Radiation at the Surface

The net downward solar energy flux can be written as the product of the downward solar flux at the surface multiplied by the absorptivity of the surface.

$$S\downarrow(0) - S\uparrow(0) = S\downarrow(0)(1 - \alpha_s)$$

The surface albedo α_s is defined as the fraction of the downward solar flux density that is reflected by the surface. Depending on the surface type and state, the surface albedo varies greatly, with values as low as 5% for oceans with mild breezes and as high as 90% for new, dry snow. Although the figures in the following table are typical, each type of surface can have a variety of albedos.

Water is the most prevalent surface, and its albedo is influenced by the solar zenith angle, cloud cover, wind speed, and water pollutants. The graph below illustrates how cloudiness and sun zenith angle affect water surface albedo.

The amount of a solar radiation beam that is reflected from a water surface depends on the angle at which the beam intersects the water. Under clear skies, the ocean's surface albedo rises sharply as the sun gets closer to the horizon. Because of how well clouds scatter radiation, sunlight falling on them no longer forms a parallel beam but is instead dispersed throughout the sky.

Under a cloud, photons that reach the surface have a roughly equal chance of coming from all possible directions, making it hard to determine the sun's position in the sky. As a result, the surface albedo under cloudy sky is not affected by the solar zenith angle. But because clouds are excellent solar radiation reflectors and their albedo is sensitive to solar zenith angle, the quantity of solar energy that reaches the surface under cloudy skies depends on it.

Table 8: *Albedos for Various Surfaces*

Surface type	Range (%)	Typical value (%)
Water		
Deep water: low wind, low altitude	5- 10	7
Deep water: high wind, high altitude	10-20	12
Bare surfaces		
Moist dark soil, high humus	5-15	10
Moist gray soil	10-20	15
Dry soil, desert	20-35	30
Wet sand	20-30	25
Dry light sand	30-40	35
Asphalt pavement	5-10	7
Concrete pavement	15-35	20
Vegetation		
Short green vegetation	10-20	17
Dry vegetation	20-30	25
Coniferous forest	10- 15	12
Deciduous forest	15-25	17
Snow and ice		
Forest with surface snow-cover	20-35	25
Sea ice, no snow-cover	25-40	30
Old, melting snow	35-65	50
Dry, cold snow	60-75	70
Fresh, dry snow	70-90	80

Source: Hartmann, D. L. (2015).

Depending on the frequency of the radiation, different surfaces have different reflectivities.

Snow and clouds reflect visible light the best, and at near-infrared wavelengths, where water absorbs a lot of energy, they become less reflective. In green plants, chlorophyll effectively absorbs radiation, resulting in an extremely low albedo for photosynthetically active radiation.

Radiation in the wavelength band from about 0.4 to 0.7 μ is effective for photosynthesis and growing plants absorb more than 90% of it.

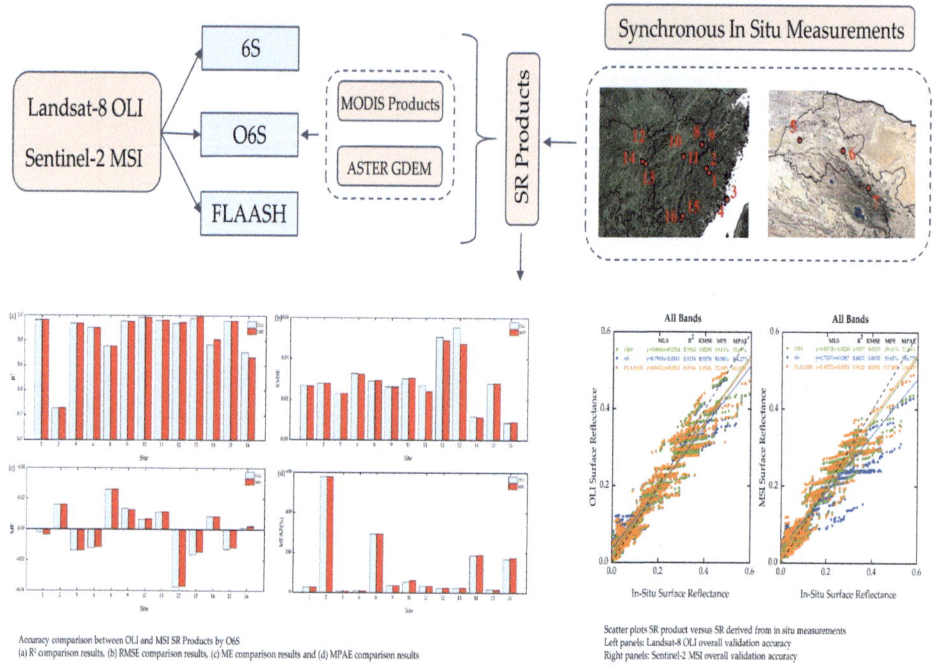

Figure 13: *Surface reflectivity for a variety of natural surfaces.*

Source: Zhou X, Liu X, Wang X, He G, Zhang Y, Wang G, Zhang Z. Evaluation of Surface Reflectance Products Based on Optimized 6S Model Using Synchronous In Situ Measurements. Remote Sensing. 2022; 14(1):83. https://doi.org/10.3390/rs14010083

For green plants, albedo for near-infrared radiation can reach as high as 50% at approximately 0.7μ, when it drastically increases. This rise in albedo is crucial for the energy budget of the surface since roughly half of the solar energy that reaches the surface is at wavelengths longer than 0.7μ. The absorption of near-infrared radiation at wavelengths longer than 0.7 μ warms the leaves without converting any of the energy to plant tissue, whereas plants require wavelengths shorter than 0.7μ for photosynthesis. The leaves stay cooler as a result of the greater albedos at wavelengths longer than 0.7μ. Green plants lose chlorophyll as they age and dry out, and their albedo at visible wavelengths rises as a result.

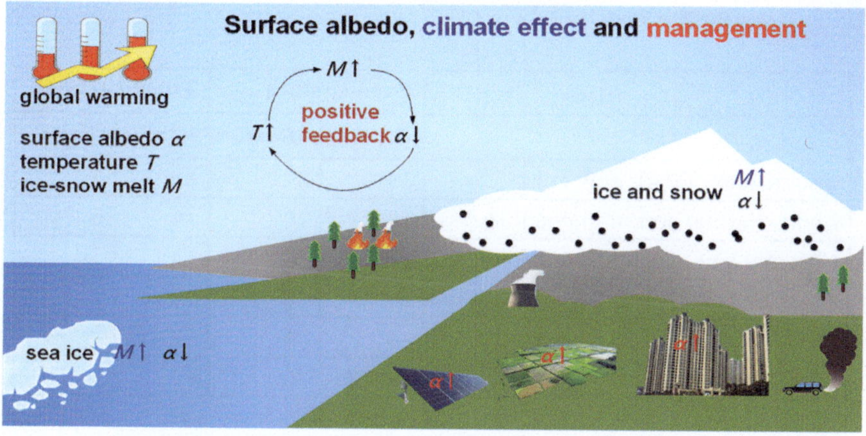

Figure 14: *The importance of surface albedo.*

Source: Zhang X, Jiao Z, Zhao C, Qu Y, Liu Q, Zhang H, Tong Y, Wang C, Li S, Guo J, et al. Review of Land Surface Albedo: Variance Characteristics, Climate Effect and Management Strategy. Remote Sensing. 2022; 14(6):1382. https://doi.org/10.3390/rs14061382

The physiological state and texture of the plant canopy influence the albedo of surfaces covered in vegetation. Albedos that are less reflective than the albedo of a single leaf can be found in leaf canopies with intricate geometries and numerous holes.

Because leaves absorb visible wavelengths more efficiently, the ratio of near-infrared to visible light declines with depth below the top of the plant canopy. Because of the increased near-infrared radiation albedo of leaves, heat from the sun can diffuse through the plant canopy and heat the soil. Blue light reflects best off of pure water.

Although various contaminants are present in natural water basins and they may reflect green light the best, their reflectance is often stronger at visible than at near-infrared wavelengths.

Near-infrared wavelengths exhibit higher reflectance in soils than visible wavelengths. Smooth soil surfaces have higher albedos than rough surfaces, and soils have much higher albedos when they are dry than when they are wet.

Table 9: *Albedos for Dry and Moist Soil Surfaces*

	Even surface		**Tilled surface**	
	Dry	***Moist***	***Dry***	***Moist***
Chernozem of dark gray color	13	8	8	4
Light chestnut soil of gray color	18	10	14	6
Chestnut soil of grayish red color	20	12	15	7
Gray sandy soil	25	18	20	11
White sand	40	20	-	-
Dark blue clay	23	16	-	-

Source: Mironova (1973).

Surface albedo can have a significant impact on surface temperature since it is very changeable and strongly influences how much solar radiation is absorbed. If surface albedo changes consistently with the climate, this can potentially have a significant impact on climate sensitivity.

Net Longwave Heating of the Surface

The downward longwave radiation from the atmosphere, the surface temperature, and the longwave emissivity of the surface must all be known in order to compute the net downward longwave radiation at the surface. The effective absorptivity of the surface equals its emissivity if the frequencies of downward longwave radiation and radiation emitted from the surface are nearly identical. This being roughly accurate, a fraction E of the downward longwave radiation at the earth is absorbed, allowing for the writing of the upward longwave at the surface;

$$F\uparrow(0) = (1-\varepsilon)F\downarrow(0) + \varepsilon\sigma T_s^4$$

Thus;

$$F\downarrow(0) - F\uparrow(0) = \varepsilon(F\downarrow(0) - \varepsilon\sigma T_s^4)$$

The downward longwave radiation from the atmosphere, the surface temperature, and the longwave emissivity of the surface must all be known in order to compute the net downward longwave radiation at the surface. The effective absorptivity of the surface equals its emissivity if the frequencies of downward longwave radiation and radiation emitted from the surface are nearly identical. This being roughly accurate, a fraction E of the downward longwave radiation at the earth is absorbed, allowing for the writing of the upward longwave at the surface;

***Table 10:** Percent Infrared Emissivity of Some Surfaces*

Water and soil surfaces		**Vegetation**	
Water	92-96	Alfalfa, dark green	95
Snow, ice granules	89	Oak leaves	91 -95
Ice	96	**Leaves and plants**	
Soil, frozen	93-94	0.8 μ	5-53
Sand, dry playa	84	1.0 μ	5-60
Sand, dry light	89-90	2.4 μ	70-97
Snow, fresh fallen	82-99.5	10.0 μ	97-98
Sand, wet	95	**Miscellaneous**	
Gravel, coarse	91-92	Paper, white	89-95
Limestone, light gray	91-92	Bricks, red	92
Concrete, dry	71-88	Plaster, white	91
Ground, moist, bare	95-98	Wood, planed oak	90
Ground, dry plowed	90	Glass pane	87-94
Natural surfaces		Paint, white	91-95
Desert	90-91	Paint, black	88-95
Grass, high dry	90	Paint, aluminum	43-55
Field and shrubs	90	Aluminum foil	1-5
Oak woodland	90	Iron, galvanized	13-28
Pine forest	90	Silver, highly polished	2
		Skin, human	95

Source: Hartmann, (2015).

The Atmospheric Boundary Layer

The lowest region of the troposphere, known as the atmospheric boundary layer, is where the influence of the surface on wind, temperature, and humidity is greatest.

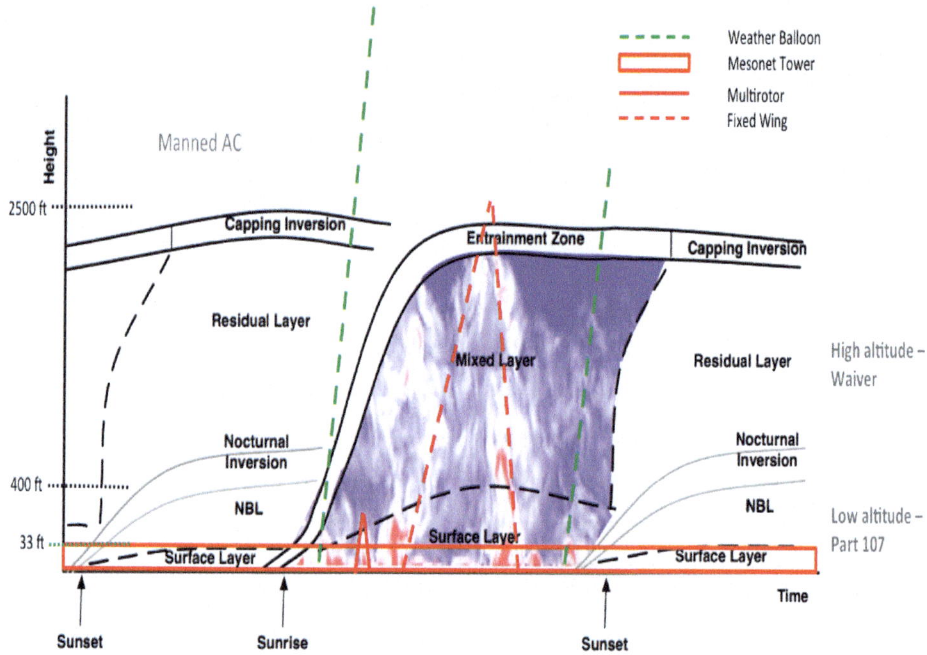

***Figure 15:** Time-height depiction of structure of the atmospheric boundary layer.*

Source: Jacob JD, Chilson PB, Houston AL, Smith SW. Considerations for Atmospheric Measurements with Small Unmanned Aircraft Systems. Atmosphere. 2018; 9(7):252. https://doi.org/10.3390/atmos9070252

- **Troposphere:** Up to a height of about 10 km, this layer is characterized by a temperature decline that occurs with height at a rate of around 6.5 oC per kilometer. This layer is the only place where all weather-related activities (water vapor, clouds, and precipitation) occur. Near the earth's surface, an aerosol layer typically exists. With a characteristic height of roughly 2 km, the aerosol concentration declines almost exponentially with height.
- **Stratosphere:** Up to an altitude of about 50 km, the temperature in the stratosphere increases with height. The temperature is roughly constant in the lower 20 km of the stratosphere. The stratopause is the principal location of ozone. Together, the troposphere and stratosphere make up more than 99 percent of the atmosphere's total mass.

- **Mesosphere:** The temperature here drops between 50 and 85 kilometers in height.
- **Thermosphere:** From roughly 85 km up to several hundred miles, this layer is present. A temperature range of 500 K to 2000 K is possible. The gases are primarily present as thin plasma, or ionized, due to solar UV radiation and energetic cosmic ray bombardment.

From its value in the free atmosphere to almost zero at the surface, the wind speed drops. Small-scale turbulent motions in the boundary layer transmit the existence of the lower boundary to the atmosphere and play a crucial role in regulating climate through fluxes of momentum, heat, and moisture. Through the atmospheric boundary layer, aerosols and gaseous chemical elements of the atmosphere are also exchanged with the surface. The atmospheric boundary layer's fast reaction to alterations in the surface environment is one of its characteristics. In the boundary layer, the response of the surface to daily variations in insolation is felt strongly, but in the free troposphere, diurnal fluctuations are often minimal, unless thermal convection related to daytime land surface heating penetrates deeply into the atmosphere. Depending on the circumstances, the depth of the atmospheric boundary layer can range from roughly 20 m to several kilometers, but a kilometer is a typical boundary-layer depth. When the surface is warm, when winds are high, when the surface is rough, and when the mean vertical motion in the free troposphere is upward, the boundary layer is typically deeper.

Turbulent motions enable the transport of mass, momentum, and energy through the boundary layer. The interaction between the surface and the atmosphere would be incredibly slow if it weren't for the chaotic swirls of turbulent motion in the boundary layer. From the depth of the boundary layer to the tiniest scales, when molecular diffusion turns into an essential transport mechanism, the turbulent motions that carry the vertical fluxes in the boundary layer range in scale. Turbulence can be produced mechanically or thermally. The mean wind in the lower atmosphere is what causes mechanical turbulence, which is produced when the mean wind is converted to turbulent motions. Warm air pockets at the surface are driven upward by their buoyancy, which causes convective turbulence. Convective turbulence occurs frequently over oceans but is more easily seen over land surfaces during the daytime when significant sun heating of the surface provides a source of buoyant energy. The boundary layer may contain a well-mixed layer where momentum, heat, and moisture are essentially invariant to height when the boundary layer is moderately unstable and buoyancy forces or shearing instabilities are producing turbulence. Depending on the weather and whether the surface is being heated or cooled, the planetary boundary layer's structure varies greatly. The boundary layer is frequently unstable and typically has a structure similar to that illustrated in the picture below when a surface heat source is available, such as over land during the day. The term "surface layer" refers to the lowest portion of the boundary layer where the vertical fluxes of moisture, heat, and momentum are almost constant with height.

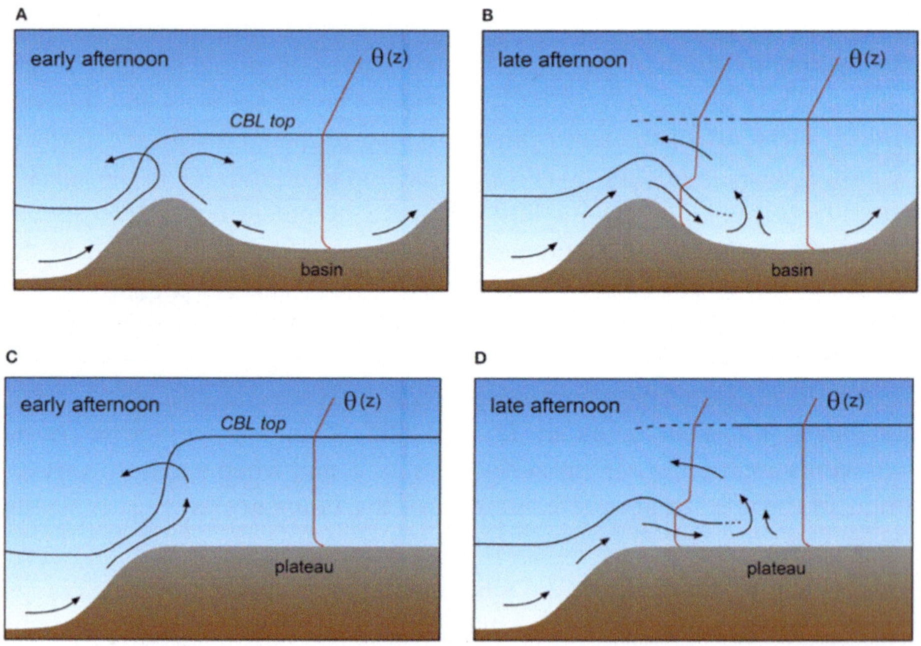

***Figure 16:** Structure of a convective boundary layer: Airflow and vertical profiles of potential temperature θ are shown by arrows and red lines, respectively.*

Source: De Wekker SFJ and Kossmann M (2015) Convective Boundary Layer Heights Over Mountainous Terrain—A Review of Concepts. Front. Earth Sci. 3:77. doi: 10.3389/feart.2015.00077

In the mixed layer, buoyancy drives turbulent motions that maintain the potential temperature Θ_v, the humidity q, and the momentum, $\dot{M}$, at values that are almost independent of height. In the mixed layer, momentum is moved downward toward the surface while heat and moisture are moved upward. The entrainment zone, also known as the top of the boundary layer, is a region where the boundary layer and free atmosphere meet. The air properties quickly shift from those of the mixed layer to those of the free atmosphere above at this transition, which is typically accompanied by a drop in humidity, an increase in potential temperature, and a reduction in the magnitude of turbulent motion-driven vertical fluxes of heat, moisture, and momentum. The process by which air from the free atmosphere is assimilated into the boundary layer is known as entrainment. A tiny downward eddy potential temperature flux in the entrainment zone is used to measure entrainment. When the boundary layer is deepening or when the large-scale flow is descending into the boundary layer, entrainment is required to contribute mass to the boundary layer. The boundary layer can become quite stable at night, trapping cold, dense air close to the surface as longwave emission cools the ground surface more quickly than the air above it. The surface becomes mechanically uncoupled from the free atmosphere and turbulence and the vertical fluxes it generates can be significantly reduced under these circumstances, while radiative transports are still possible. Since the transfer of potential temperature is downhill and the potential temperature rises rapidly towards the surface, less dense air is being driven downward against buoyancy. The mean wind speed shear, which under these circumstances frequently has a low-level wind maximum near the top of the boundary layer and weak winds near the surface, supplies the energy for mixing less dense air downward toward the surface.

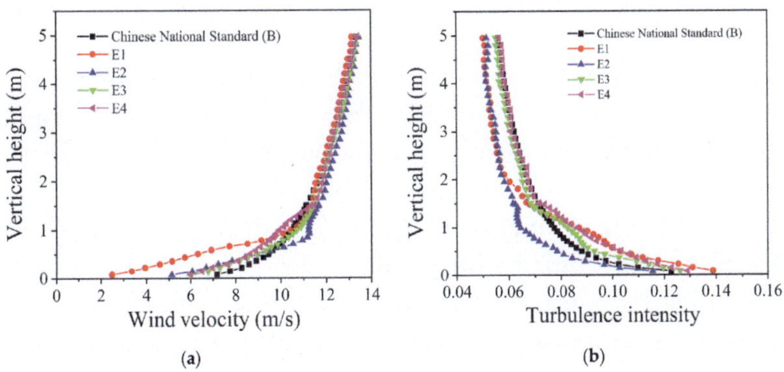

***Figure 17:** Wind velocity and turbulence intensity profiles at different* x *positions on the central axis of the model region: **(a)** Wind velocity profile; **(b)** Turbulence intensity profile.*

Source: Chen Z, Wei C, Chen Z, Wang S, Tang L. Numerical Simulation of Atmospheric Boundary Layer Turbulence in a Wind Tunnel Based on a Hybrid Method. Atmosphere. 2022; 13(12):2044. https://doi.org/10.3390/atmos13122044

Thus, when the wind speed in the open atmosphere is light and offers little energy for mixing warm air downward to the radiatively cooled surface, the minimum surface air temperature obtained on a clear night is often lower. Clouds that are crucial to boundary-layer physics and vertical transports can be found in the atmospheric boundary layer. The release of latent heat in clouds can give the boundary layer buoyancy to propel vertical motions. Fair-weather cumulus clouds and stratocumulus clouds are examples of boundary layer clouds that are crucial to climate. Fog is a boundary-layer cloud that is also significant, despite being less common. Since the high potential temperature air that drives deep convection is produced in the boundary layer, the boundary layer also interacts in significant ways with deep convective clouds. The tops of boundary layer clouds typically occur towards the top of the boundary layer and their bases are a few miles above the surface, with the exception of when fog is present, meaning that a cloud and a sub-cloud layer exist within the boundary layer. Through their convective heat and moisture fluxes as well as through their radiative effects, stratocumulus clouds alter the boundary-layer physics. Longwave cooling from stratocumulus cloud tops can be a significant mechanism for producing buoyancy within the boundary layer because they emit longwave radiation effectively and are relatively warm. When the air at the top of the boundary layer is cooled, it sinks and is eventually replaced by warmer pockets of air rising from below. The graph below depicts the temperature changes throughout the day in the lowest 1500 m of the atmosphere above Nebraska on a typically clear summer day. At daybreak, the air one kilometer above the surface is cooler than the surface itself. After sunrise, this temperature inversion soon vanishes as insolation warms the surface and transfers this heat to a shallow layer of air close to the ground. The surface reaches its peak temperature about midday, and a lapse rate close to the dry adiabatic value of 9.8 K km-1 is seen close to the surface. Warm air pockets are raised near the surface at this time by buoyancy, and the boundary layer's turbulent convection effectively transports sensible heat upward. As a result of the efficient upward transmission of energy caused by turbulent oscillations, the surface starts to cool even before dusk.

After sunset, the surface rapidly cools until it reaches its nighttime temperature by 10 PM, leaving a very pronounced inversion close to the ground. The flow of heat, moisture, and momentum between the atmosphere and the surface will be impacted by tidal fluctuations in the temperature profile when mean wind speeds are light to moderate. The turbulent movement of momentum from the free atmosphere is suppressed by the substantial density stratification brought on by a nighttime temperature inversion. Strong density stratification also suppresses sensible and latent heat fluxes, thus the surface temperature is predominantly affected by radiative forcing, which acts to chill the surface at night. When a high-pressure system dominates the local weather pattern, temperature inversions frequently form. Strong inversions are made possible by the weak surface winds, clear sky, and downward mean motion in the free atmosphere that are typically associated with high-pressure systems. Because turbulent mixing into the free atmosphere is inhibited by the large density gradient, pollutants emitted during a temperature inversion can readily accumulate to dangerous levels.

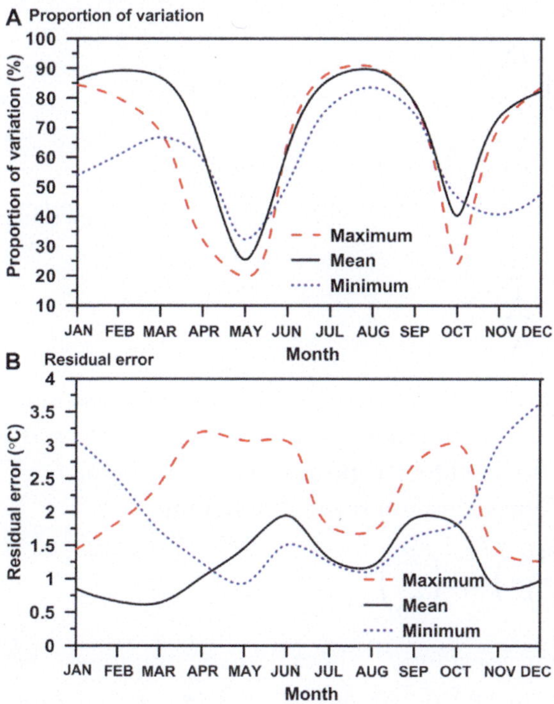

Figure 18: *Annual variation in the proportion of variation (A) and residual error (B) resulting from multi-coolinearity assessment in the Sudanese desert.*

Source: Kattel DB, Salih HAM, Yao T and Ullah K (2022) Near-Surface Air Temperature Dependence on Elevation and Geographical Coordinates Over Tropical Desert Land Surfaces. Front. Earth Sci. 9:777381. doi: 10.3389/feart.2021.777381

The Neutral Boundary Layer

Buoyancy does not play a significant role in the turbulent kinetic energy budget when the static stability in the boundary layer is close to neutral. The kinetic energy of the free atmosphere average wind serves as the source of energy for boundary-layer turbulence in neutral conditions. A significant flux of momentum to the surface is produced by the turbulence in the boundary layer. The horizontal momentum at the surface that is being transferred vertically, or 0, acts as a drag on the air flow. In the surface layer the vertical gradient of wind speed (U) under conditions of neutral stability should depend only on the height (z), density (ρ), and surface drag. The characteristic wind speed for use in dimensional analysis is the friction velocity, u*.

$$u_* = \left(\frac{\tau_0}{\rho}\right)^{1/2}$$

Dimensional analysis implies that the scaled wind shear should be a constant when the friction velocity and height are used to scale the wind shear.

$$\left(\frac{z}{u_*}\right)\frac{\partial u}{\partial z} = \frac{1}{K}$$

Regardless of the surface properties, the von Karman constant, K, is constant for all neutral boundary layers and has a measured value of roughly 0.4. The logarithmic velocity profile can be obtained by integrating the aforementioned equation with respect to height.

$$U(z) = \left(\frac{u_*}{k}\right) ln\left(\frac{z}{z_0}\right)$$

An additional constant Z_0, is called the roughness height; the height at which the wind speed reaches zero. For most natural surfaces, their regularities of the surface are larger than the 1 mm depth of the layer where molecular diffusion dominates, and this roughness can be characterized by the height Z_0. Roughness heights for natural surfaces range from about a millimeter for average seas to more than a meter for cities with tall buildings. Roughness heights are estimated by measuring the wind speed profile under neutral conditions and solving for Z_0. The logarithmic velocity profile has been shown to be a good approximation for many laboratory boundary layers and also for the planetary boundary layer under conditions of neutral stratification. It is valid for heights much greater than the roughness height, ($Z >> Z_0$), and so does not describe the mean wind velocity profile within the plant canopy or very close to a rough surface. The logarithmic velocity profile law is useful for expressing the momentum flux at the surface in terms of the wind speed at some height in the surface layer. Substituting u* into U(z) yields an expression for the surface drag in terms of the wind speed, U_r, at some reference height z_r.

$$\tau_0 = \rho C_D U_r^2$$

Where;

$$C_D = k^2 \ln \left(\frac{z_r}{z_0}\right)^{-2}$$

The drag coefficient, C_D depends on the ratio of the reference height to the roughness height. The reference layer can be determined at any depth inside the surface layer where measurements are practical to take and the logarithmic profile accurately represents the flow. The empirical estimates of surface fluxes and the specification of surface fluxes in climate models are both based on the aerodynamic drag formula and associated formulas for the sensible and latent heat fluxes at the surface. They enable turbulent fluxes to be calculated with just the mean wind speed at a reference height and a few more variables.

Stratified Boundary Layers

Stratified boundary layers can be included in the dimensional analysis for neutral boundary layers2. In this theory, buoyancy and heat flux variables are added to the dimensional analysis.

The vertical coordinate is scaled by a dimensionless combination of the friction velocity, the heat flux, and the buoyancy in order to create characteristic vertical profiles for wind and temperature. These profiles can be used to generate bulk aerodynamic formulas that express the turbulent heat and momentum fluxes at the surface as mean variables. These formulae coefficients now take into account both the atmosphere's vertical stability and the height of its roughness. Only the top layer is covered by this idea. The Richardson number can be used to describe the vertical stability. In differential form the Richardson number depends on the vertical derivatives of potential temperature Θ, and wind speed, U;

$$Ri = \frac{g}{T_0} \frac{(\partial \Theta / \partial z)}{(\partial U / \partial z)^2}$$

where g is the gravitational acceleration and (T_0) is the reference temperature. The bulk Richardson number for the boundary layer may be written as;

$$Ri = \frac{g}{T_0} \left[\frac{z_r \left(\Theta(z_r) - \Theta(z_0) \right)}{U(z_r)^2} \right]$$

When the potential temperature of the air at the surface is higher than the potential temperature at the surface, the Richardson number is high. Such circumstances result in stable stratification of the air, which prevents vertical mixing. When parcels are lifted off the surface, they become negatively buoyant and are pulled back down by gravity. The kinetic energy present in the mean wind shear close to the ground, which can produce turbulent velocities sufficient to mix stabilized stratified air, can counteract this stabilizing effect. The square of the wind speed in the denominator of represents this kinetic energy contribution. Small vertical displacements of parcels will be accelerated by the buoyancy force if the potential temperature at the surface falls off with height. This indicates that the boundary layer is unstable. A buoyantly unstable boundary layer has a negative Richardson number. Due to the easy exchange of pockets over flat surfaces, vertical energy and moisture transmission is relatively efficient under these circumstances. In terrestrial areas, the boundary layer frequently becomes unstable during the summer as insolation warms the surface. A temperature inversion forms at night when the surface cools more quickly than the air above it. Strong density stratification that results from this can reduce nighttime heat, momentum, and moisture exchanges between the surface and free atmosphere. The diurnal variations in the wind profile can be used to observe how changes in the stability of the boundary layer affect momentum flows. Due to the lower downward mixing of momentum from the free atmosphere caused by the nocturnal increased static stability, wind speeds measured very close to the surface diminish. Because there is less surface drag at night, the winds higher in the boundary layer are stronger. The wind speed near the surface increases during the day at the expense of the wind speed higher in the boundary layer due to efficient mixing of momentum through a rather unstable boundary layer.

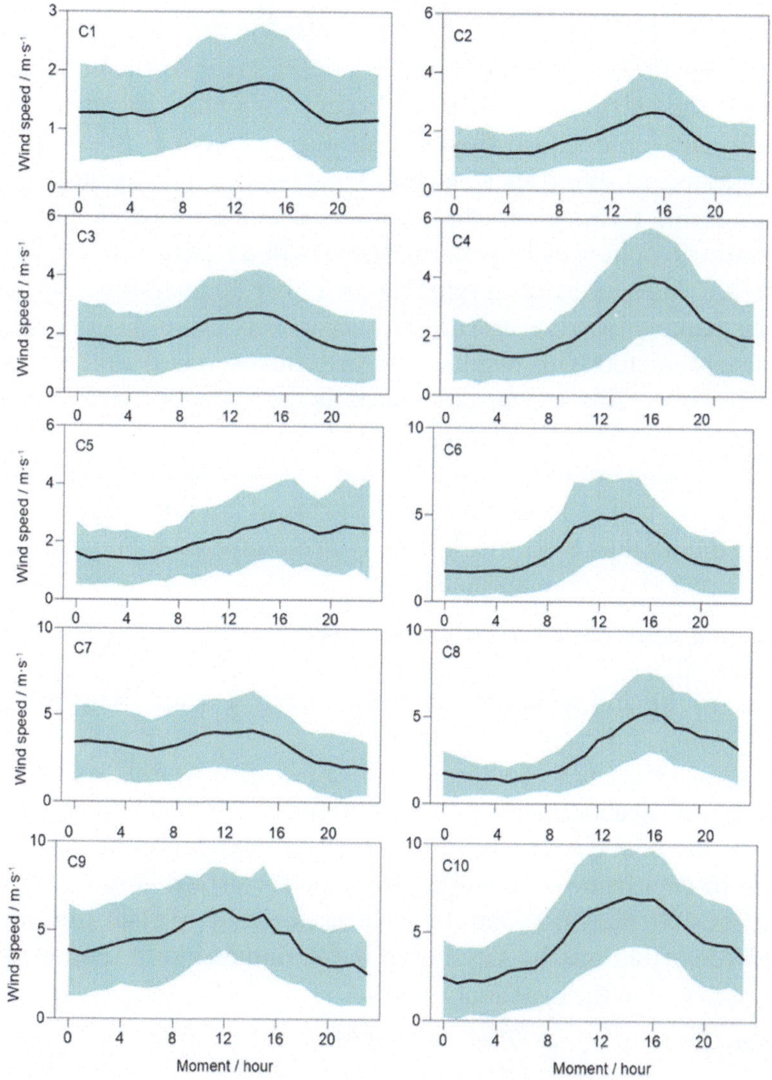

***Figure 19**: The cluster analysis results of the hourly wind speed in 3,650 days at Shunyi station in Beijing. The* X*-axis is 00:00–23:00 CST, the* Y*-axis is the wind speed, and C1–C10 represent the 10 clusters, respectively. The black line represents the sample average at different times, and the shade represents one standard deviation of the corresponding sample.*

Source: Yan P, Zuo D, Yang P and Li S (2021) Typical Modes of the Wind Speed Diurnal Variation in Beijing Based on the Clustering Method. Front. Phys. 9:675922. doi: 10.3389/fphy.2021.675922

Sensible and Latent Heat Fluxes in the Boundary Layer

The turbulent fluid dynamics in the boundary layer result in sensible and latent heat fluxes from the surface. Except for within about a millimeter of the surface, molecular diffusion transport is minimal in comparison to turbulent transport. Wind speed changes that happen quickly and chaotically are what define turbulence. The turbulent changes in wind speed will be accompanied by changes in scalar properties like temperature and humidity if there are mean vertical gradients in either temperature or humidity. When air parcels flowing upward and downward have distinct properties, turbulence is created. This results in vertical fluxes of mass, momentum, and energy. Thus, the flux can be calculated as the spatial or temporal average of the product of the interest attribute and vertical velocity. For instance, if we know the temperature, T, and vertical velocity, w, at a location close to the surface, we may calculate the vertical flux of sensible heat by averaging over time the product of vertical velocity and temperature, multiplied by the specific heat and average air density.

$$Upward\ sensible\ heat\ flux = C_p\ \rho\ \overline{wT}$$

Temperature and wind measurements must be made frequently enough to identify the turbulent oscillations that cause the vertical transport in order for this estimate to be correct. Since turbulent variations happen very quickly, direct measurements of turbulent fluxes may need to be made more frequently than once every second. It may be challenging to derive typical fluxes from data made at a single place due to surface homogeneities. The upward sensible heat flow can be expressed as the total of the time mean and eddy contributions by separating the variables into a time mean and a departure from the time mean, or eddy.

$$w = \overline{w} + w' , \qquad T = \overline{T} + T'$$

Here, an overbar denotes the quantity under the overbar's time average, and a prime denotes a departure from that time average. Using the aforementioned equations to substitute and perform averages, we can determine the mean and eddy contributions to the vertical flux of temperature.

$$\overline{wT} = \overline{w}\overline{T} + \overline{w'T'}$$

Total = mean+ eddy

In comparison to typical eddy or turbulent vertical velocities at the surface, the mean vertical velocity is quite low, and the vertical heat flux is mostly contributed by eddies. The eddy fluxes of heat and moisture at a certain level in the atmospheric boundary layer can then be used to define the latent and sensible heat fluxes.

$$SH = C_p \, \rho \, \overline{wT}, \qquad LE = L \, \rho \, \overline{wq}$$

where ρ is the air density, C_p is the specific heat of air at constant pressure, and L is the latent heat of vaporization. It is not common practice to measure the turbulent velocity, temperature, and moisture variations required to determine the sensible and latent energy flows and global climate models do not mimic these quick, small-scale oscillations. Most often, one must use variables averaged over wider geographical and temporal scales than those of the turbulent motions in the boundary layer in order to calculate the turbulent fluxes. Surface fluxes can be estimated using a variety of techniques using mean variable measurements. The most popular approach is to connect the turbulent fluxes to observable spatial or temporal averages using bulk aerodynamic formulae. One might hypothesize that the sensible heat flux is proportional to the temperature difference between the surface and the air at some standard altitude, z_r, where mean variables are known. Since some of the kinetic energy of boundary layer turbulence comes from the mean winds blowing over the surface, we mightassume that the turbulent fluxes are also proportional to the mean wind speed, U_r, at the standard height. We establish an expression that connects the sensible heat flux to the mean wind speed and temperatures using the similarity theory, and these fundamental presumptions are consistent with the theory's findings.

$$SH = C_p \, \rho \, C_{DH} U_r \left(T_s - T_a(z_r)\right)$$

The difference in specific humidity, q, between the surface and the atmosphere at the reference height can be used to determine the latent heat flux.

$$LE = L \, \rho \, C_{DE} U_r \left(q_s - q_a(z_r)\right)$$

In the bulk the above aerodynamic formulas for SH and LEρ is the air density, C_p is the specific heat at constant pressure, and C_{DH} and C_{DE} are aerodynamic transfer coefficients for temperature and humidity, respectively. Subscripts s and a indicate values for the surface and the air at the reference level, respectively. The aerodynamic transfer coefficients depend on the surface roughness, the bulk Richardson number, and the reference height. Under ordinary circumstances the values of the transfer coefficients for heat, moisture, and momentum would be nearly equal, and typical values for neutral stability and for a 10-m height abovethe surface would range from 1x 10^{-3}over the ocean to 4 x10^{-3} over moderately rough land. If the wind speed at 10 m is 5 m s^{-1} and C_D = 3 x 10^{-3}then from the SH equation above the flux of sensible heat across the surface layer of the atmosphere is about 15 W m^{-2} for each degree of temperature difference between the surface and the air at 10 m.

Equilibrium Bowen Ratio for Saturated Conditions

Because of the relationship between temperature and saturation vapor pressure, the latent heat flux is sensitively temperature dependent. Over water or wet land surfaces we may assume that the mixing ratio of water vapor at the surface is equal to the saturation mixing ratio, q*, at the temperature of the surface.

$$q_s = q^*(T_s)$$

A first-order Taylor series can be used to estimate the saturation vapor mixing ratio of the air at the reference height.

$$q_a^* = q_s^*(T_s) + \frac{\partial q^*}{\partial T} \cdot (T_a - T_s) + \cdots$$

The relative humidity at that altitude can be used to indicate the actual vapor mixing ratio of the air there.

$$RH = \frac{q}{q^*}$$

$$q_a \cong RH \cdot \left(q_s^*(T_s) + \frac{\partial q^*}{\partial T} \cdot (T_a + T_s) \right)$$

Substituting (q_a) into the LE equation above yields an expression for the heat loss from the surface through evaporation in terms of the temperature difference and the relative humidity,

$$LE = L\,\rho\, C_{DE} U_r (q_s - q_a(z_r)) \left(q_s^*(1 - RH) + RH\, B_e^{-1} \frac{C_p}{L} (T_s + T_a) \right)$$

Where;

$$B_e^{-1} \equiv \frac{C_p}{L} \frac{\partial q^*}{\partial T}$$

The sensible cooling to surface latent cooling ratio is known as the Bowen ratio. By comparing the LE and SH equations, we can observe that the Bowen ratio takes on a certain value when the surface is moist and the air is saturated, or when RH = 1:

$$B_0 \equiv \frac{SH}{LE} = B_e$$

When the surface and the air at the reference level are saturated, the Bowen ratio approaches the value Be given by (B_e^{-1}), which can be called the equilibrium Bowen ratio.

In order for the humidity at the reference height to be in equilibrium at the saturation level, we assume that the flux of moisture from the boundary layer to the free atmosphere is just enough to balance the upward flux of moisture from the surface.

In such an equilibrium, the rate of change of the saturation mixing ratio of water vapor with temperature is inversely proportional to the Bowen ratio. The saturation mixing ratio's rate of change with temperature is highly dependent on the ambient temperature. Also;

$$\frac{\partial q^*}{\partial T} \approx q^*(T) \cdot \left(\frac{L}{R_V T^2}\right)$$

The equilibrium Bowen ratio drops exponentially with temperature because the exponential dependency of the saturation mixing ratio on temperature surpasses the inverse square of temperature. In the graph, a log-linear plot depicts the saturation mixing ratio and equilibrium Bowen ratio's temperature relationships.

At approximately 0°C, the equilibrium Bowen ratio is unity; at around 300C, it is approximately 0.2. The evaporative cooling increases as the relative humidity decreases from 1.0 to decreasing values, making the equilibrium Bowen ratio the highest attainable Bowen ratio for a wet surface.

Since the air at the reference height is typically not saturated, the actual Bowen ratio over a wet surface will typically be lower than the equilibrium Bowen ratio. Latent cooling of the surface predominates sensible cooling from a wet surface at temperatures like those in the tropics due to the significant temperature dependency of saturation vapor pressure, but in high latitudes during the winter, sensible heat transport can be of most significance.

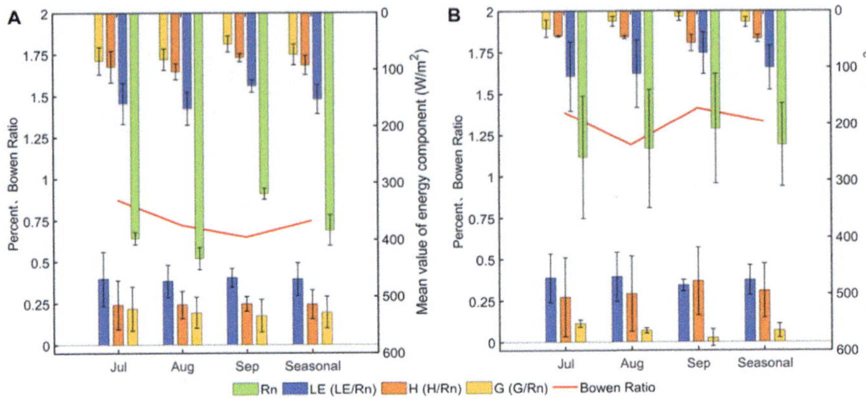

Figure 20: *Monthly and seasonal averages of the energy components, ratios to net radiation, and Bowen ratios in the (A) western and (B) eastern regions.* R_n *is the net radiation,* LE *is the latent heat flux,* H *is the sensible heat flux, and* G *is the soil heat flux.*

Source: Ren X, Zhang Q, Yue P, Yan X and Yang Y (2022) Energy Distribution Characteristics and Environmental Impact Factors in Different Regions of the Summer Monsoon Transition Zone. Front. Environ. Sci. 10:847725. doi: 10.3389/fenvs.2022.847725 s

The discussion that came before strictly only applies to situations where the surface is wet, preventing evaporative cooling from being restricted by a lack of surface moisture. When moisture cannot be provided from below the surface quickly enough to keep the air in contact with the surface saturated, evaporative cooling across land expanses may be significantly diminished. Sensible cooling and longwave emission are required to counterbalance solar heating in desert regions since the surface is typically so dry that evaporative cooling is minimal regardless of the temperature. The physical and biological state of the plant canopy and the water content of the soil regulate cooling via evaporation and transpiration through leaves for vegetated terrain.

VARIATION OF ENERGY BALANCE COMPONENTS WITH LATITUDE

Their dependency on latitude provides some insight into the relative importance of the different elements of the surface energy budget as a function of the meteorological regime.

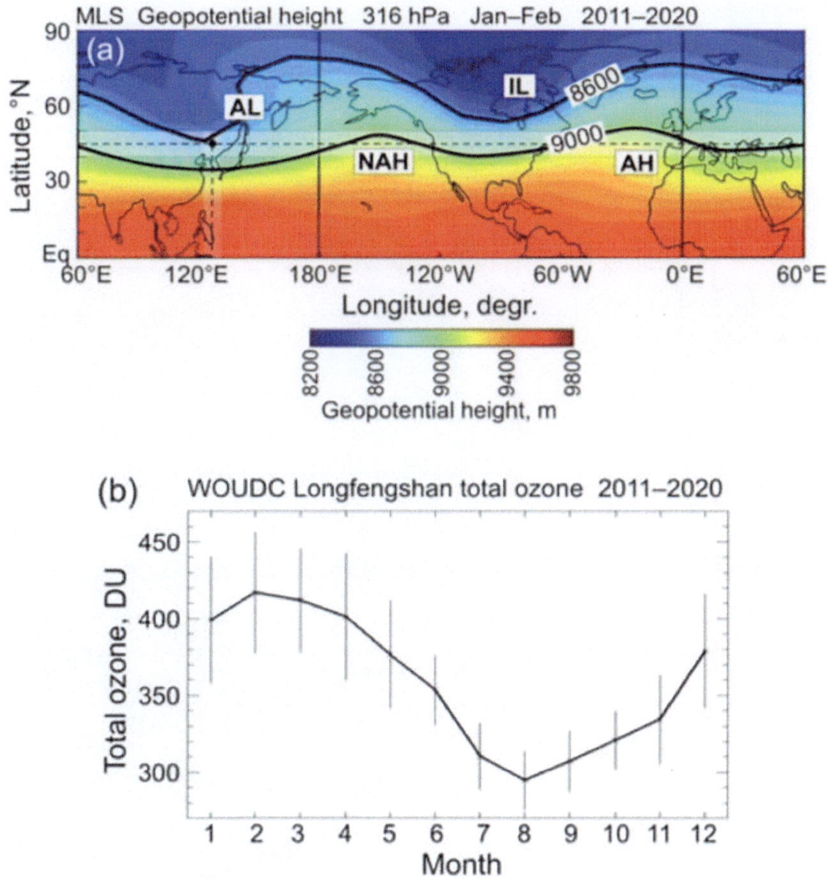

Figure 21: *Components of the annual-average surface energy balance plotted against latitude.*

Source: Zhang C, Evtushevsky O, Milinevsky G, Klekociuk A, Andrienko Y, Shulga V, Han W, Shi Y. The Annual Cycle in Mid-Latitude Stratospheric and Mesospheric Ozone Associated with Quasi-Stationary Wave Structure by the MLS Data 2011–2020. Remote Sensing. 2022; 14(10):2309. https://doi.org/10.3390/rs14102309Following the general pattern of insolation at the top of the atmosphere, the net radiation at the surface reaches its maximum in the tropics. At the same latitudes, the net radiation over the tropical water is roughly 60% higher than it is over the land. Surface albedos on land are often higher than on water surfaces. The larger net radiation over the oceans is mostly caused by low surface albedo, although changes in surface temperature and water vapor abundance between land and ocean regions also play a role.

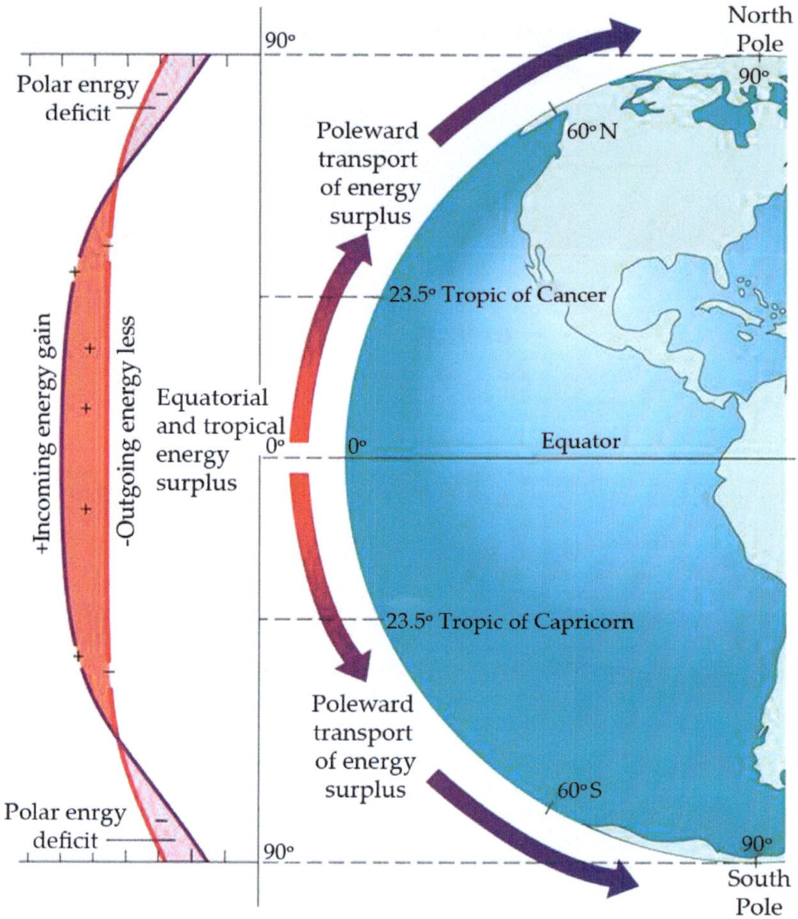

***Figure 22:** Planetary Energy Balance.*

Source: https://www.jkgeography.com/

In polar regions, a downward flux of sensible heat balances out the net radiative energy loss from the surface. In high latitudes, the sensible heat flux downward is linked to the temperature inversion. The poleward heat movement in the atmosphere, which maintains the polar areas warmer than they would be in the absence of atmospheric transport, supports the polar temperature inversion. Evaporation balances out 90% of the worldwide ocean surface's radiative heating. In the subtropics over seas, where surface solar heating and surface water temperature are both at their highest, the cooling of the surface via evaporation is greatest. Additionally, the trade winds provide a source of kinetic energy for turbulence in the boundary layer, and mean downward motion in the free troposphere of the subtropics pulls dry air into the boundary layer, increasing evaporation. Due to the dryness of the continents, evaporation on land peaks close to the equator and decreases in the latitude band from 10 to 30 degrees. Although there is energy available to evaporate water over the subtropical land masses, there is not much surface water available for evaporation. In the subtropics, where hot, arid conditions are most common, the sensible heat loss from the continent's summits rises. Over both oceans and land, the Bowen ratio—the ratio of sensible to latent cooling of the surface—increases with latitude. This is in line with the preceding section's explanation of how the equilibrium Bowen ratio rises with decreasing temperature due to the strong temperature dependency of saturation vapor pressure. In middle and high latitudes, the energy provided by radiative heating is equivalent to or greater than the energy provided by evaporative cooling from the surface. Heat is transported significantly from the tropics to middle and high latitudes by ocean currents. The extremely significant evaporative heat loss in middle latitudes is supported by this lateral energy movement in the ocean. The Bowen ratio allows for a simple classification of the continents.

Table 11: *Classification of surface based on Bowen ratio*

Type of surface	Range of Bowen ratios
Deserts	>10.0
Semi-arid landscapes	2.0-6.0
Temperate forests and grasslands	0.4-0.8
Tropical rainforests	0.1-0.3
Tropical oceans	<0.1

Source: Lewis, J. M. (1995).

The majority of "desert" continents are those with high Bowen ratios, while "wet" continents are those with low Bowen ratios. With a Bowen ratio of roughly 2, Australia is without a doubt the driest continent, followed by Africa, which is home to the Sahara and Kalahari deserts. With Bowen ratios of roughly 0.6 each, South America and Europe are the wettest continents. With a Bowen ratio well below 1.0, North America is a similarly moist continent. Averaged over all land areas, the sensible and latent surface cooling rates are roughly identical, but over the world's oceans, the sensible cooling rate is only approximately one-tenth that of the latent cooling.

Table 12: Annual Energy Balance of the Oceans and Continents

Area	R_s	LE	SH	ΔF_{eo}	SH/LH
Europe	52	32	20	0	0.62
Asia	62	29	33	0	1.14
North America	53	30	22	0	0.74
South America	93	60	33	0	0.56
Africa	90	34	56	0	1.61
Australia	93	29	64	0	2.18
Antarctica	-16	0	-15	0	-
All land	**65**	**33**	**32**	**0**	**0.96**
Atlantic Ocean	I09	95	1 1	3	0.11
Indian Ocean	113	I02	9	1	0.09
Pacific Ocean	114	103	11	0	0.10
Arctic Ocean	-5	7	-7	-5	-1.00
All oceans	**I09**	**98**	**11**	**0**	**0.11**

Source: Budyko (1963)

Diurnal Variation of the Surface Energy Balance

The diurnal change of insolation has a significant impact on the surface energy balance everywhere but the polar regions.

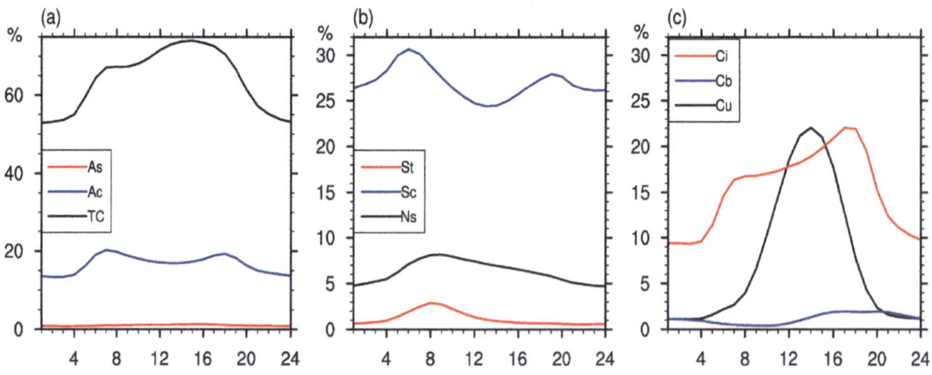

***Figure 23:** Diurnal variation of the surface.*

*The figure shows the diurnal variation of the occurrence frequencies (%) of various cloud types in central and east China during 1981–2011: **(a)** TC, Ac, As, **(b)** Ns, Sc, St, **(c)** Cu, Cb, Ci(the horizontal axis represents the local so-*

lar time, and the vertical axis represents the occurrence frequency of clouds) The dataset includes the information of total cloud (TC), and 10 cloud genera (cirrus (Ci), cirrocumulus (Cc), cirrostratus (Cs), altocumulus (Ac), altostratus (As), nimbostratus (Ns), stratocumulus (Sc), stratus (St), cumulus (Cu), and cumulonimbus (Cb)).

Source: Gao C, Li Y, Chen H. Diurnal Variations of Different Cloud Types and the Relationship between the Diurnal Variations of Clouds and Precipitation in Central and East China. Atmosphere. 2019; 10(6):304. https://doi.org/10.3390/atmos10060304

The surface is covered in a thick layer of both living and dead grass, and the albedo is only approximately 16 percent. The 24-hour period displayed had an average net radiation of 155 Wm^{-1}, with 263 Wm^{-2} coming from the sun and 108 Wm^{-2} coming from infrared fluxes. Near local solar noon, the net downward solar radiation at the earth reaches its highest at about 700 Wm^{-2}. Due to the great insolation, lack of clouds, and relatively low surface albedo, the daily solar heating is significant. About 300 Wm^{-2} of downward longwave radiation is emitted daily, with little variation. Due to the negligible diurnal fluctuation in air temperature in the free atmosphere, the downward longwave radiation varies very little throughout the day. Prior to sunrise, the surface upward emission is roughly 350Wm^{-2} and increases to approximately 500Wm^{-1} by midday in response to the higher surface temperatures during the day. Prior to daybreak, the surface temperature ranges from around 10°C to roughly 40°C by midday. This results in an increase in the net longwave loss from the surface from around 50 Wm^{-2} at night to about 200 Wm^{-2} at noon. Given the bright skies and low air humidity, the longwave loss is comparatively high. The net radiation as a result is a substantial solar-driven gain peaking at around 500 Wm_{-2} at noon, and an essentially uniform loss of 50 Wm^{-2} by longwave cooling during the night.

For example, in the diurnal cycle in the surface energy balance elements for a dry, desert lake bed, evaporation has almost no part in the energy balance because the surface soil moisture at the time the observations were conducted was only approximately 2%. Since the surface emits radiation into the arid desert air at night, the net radiation balance is negative. The energy lost from the soil as it cools throughout the night balances the radiative cooling of the surface. Due to minimal winds and the fact that the cold air at the surface is denser than the warmer air above, the sensible heat flux is almost zero at night. The nighttime temperature inversion's great vertical stability inhibits turbulence in the energy exchange. Around five in the morning, the sun rises, and the surface cooling quickly subsides. The ground's ability to store heat helps to offset the net radiative heating in the morning. After midday, air turbulence plays a major role in balancing out the intense daily radiative heating. Soon after midday, the sensible heat flux reaches its maximum, and by mid-afternoon, the sensible cooling has surpassed the net radiation, causing the surface to start cooling.

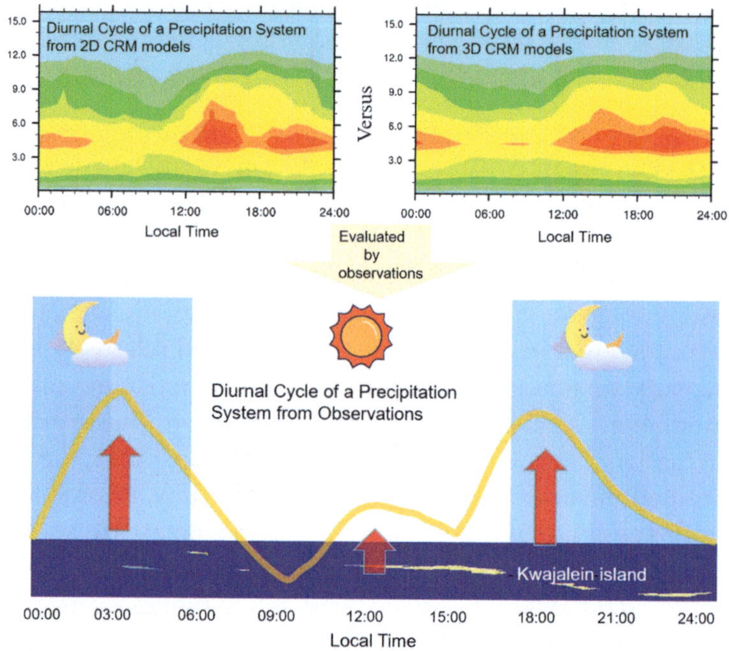

Figure 24: *Diurnal cycle from the Tropical Rainfall Measuring Mission Kwajalein Experiment (KWAJEX).*

Source: Xu H, Song Y, Hu T, Wang J, Zhang D. Numerical Simulation of the Diurnal Cycle of a Precipitation System during KWAJEX by 2D and 3D Cloud-Resolving Models. Remote Sensing. 2022; 14(23):5955. https://doi.org/10.3390/rs14235955

Seasonal Variation of the Energy Balance of Land Areas

Seasonal variations in insolation and temperature cause the surface's energy balance to shift, particularly in middle and high latitudes. Even when the temperature and insolation stay largely constant, the energy balance in tropical areas may fluctuate due to seasonal variations in precipitation. The figure below depicts the seasonal change of the surface energy balance's constituent parts at several middle latitude regions. Depending on the latitude, sky conditions, and surface albedo, the annual variation in net radiation often follows that of insolation, with peak values in the summer approaching 200 Wm-2 in land areas. Summer net radiation on water surfaces can reach 300 Wm-2 in locations with few clouds. The local surface conditions determine the technique for balancing this net radiation. Since storage is minimal and transit is nonexistent over land expanses, the main concern is whether the mechanism is latent or perceptible cooling. The distribution of surface moisture, air temperature, and air humidity all affect how much cooling is sensible and latent. The latent cooling is often constrained by and follows the annual cycle of radiative heating in places with high summer precipitation, when the surface remains relatively moist. The latent cooling is minimal in regions with particularly dry climates, such as Yuma, Arizona, unless it rains during those months. At Yuma, there is a lot of evaporation in the spring and in September. Since Flagstaff, Arizona, is higher up than Yuma, there is more summertime precipitation in the mountains. As a result, there is more evaporative cooling here throughout the summer than there is in Yuma.

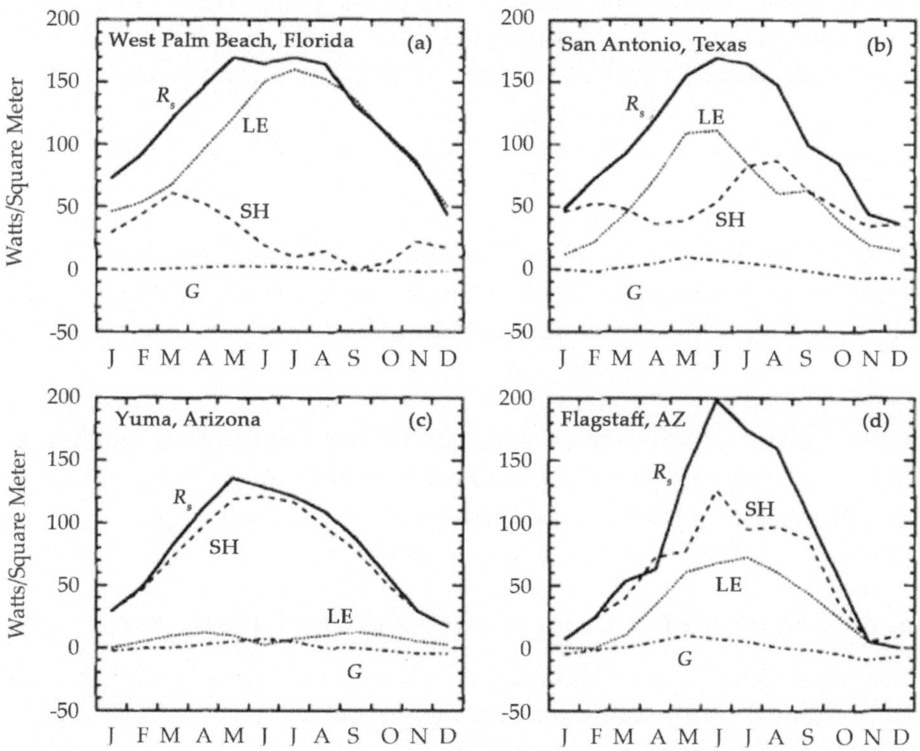

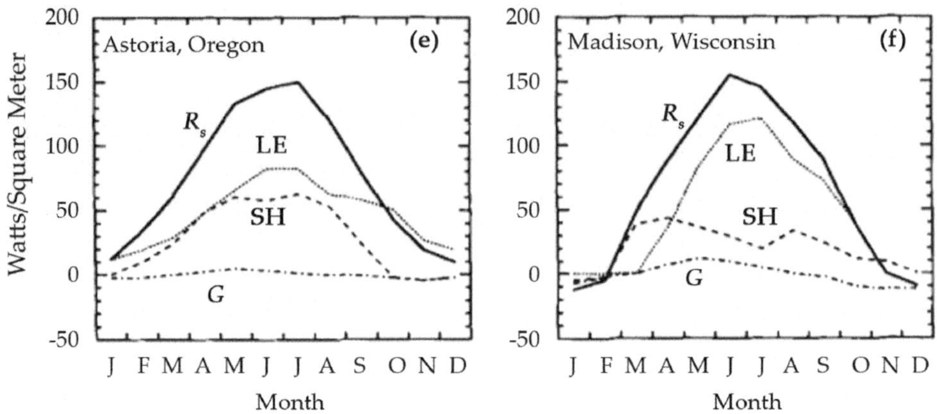

Figure 25: *Annual cycle of heat budget components for various midlatitude land locations.*

Source: Sellers (1965)

Because the winters are relatively dry in West Palm Beach, Florida, the springtime insolation increase is initially balanced by equal contributions from the surface's latent and sensible cooling. Evaporative cooling takes hold as the convective precipitation of summer starts to soak the surface. At San Antonio, Texas, the surface dries out in the spring and summer, which causes the significance of sensible heating to gradually increase over the summer. Even though Astoria, Oregon receives more annual precipitation than Madison, Wisconsin, where evaporative cooling predominates in the summer, sensible cooling is significant there. This is due to the fact that Astoria experiences its highest precipitation during the winter, when there is little energy available for evaporation. As a result, around 70% of the city's yearly precipitation rushes off before it can evaporate. Astoria has comparatively little precipitation during the summer. At Madison, summertime sees a surge in precipitation as a result of the high net radiation, which gives off enough energy to release significant volumes of water through evaporation into convective weather systems like thunderstorms.

Surface Energy Flux Components over the Oceans

In oceanic regions, the water has a high enough heat capacity that the energy needed for evaporation can come directly from the water's energy. In comparison to variables like wind speed or the temperature and humidity contrast between the surface and the air above it, the evaporative loss may be less associated with net radiation. The Kuroshio and Gulf Stream currents have a major role in the wintertime evaporation that occurs over the oceans at high latitudes. Warm water is carried poleward by these currents along the western edges of the Pacific and Atlantic Oceans, where it meets dry, chilly air coming off the continents. In certain areas during the winter, the combination results in evaporation rates of around 400 W m^{-2}.

CHAPTER 4

Soil and Its Heat Balance

CONTENTS

OVERVIEW

Agroclimatology is a special study of climatology that focuses on terrestrial, near-surface processes as they relate to the production of crops, livestock, and forests. The relationship between large-scale meteorological phenomena including air and ocean currents and surface conditions as impacted by elevation, latitude, slope aspect, and vegetation cover determines the climate close to the ground. These climatic conditions have a significant impact on crop growth and output, and energy exchange inside and above the crop canopy affects the local microclimate. The distribution of incoming energy between warming the soil, evaporating water from the soil and plants, and warming the air above the ground surface is influenced by soil qualities such as water content, color, texture, and density.

Climatic conditions on the surface of the planet are influenced by large-scale meteorological processes including air and ocean currents as well as variations in physical position (elevation, latitude, and aspect). These climatic conditions have a direct impact on the structure and density of plant canopies in natural systems. In contrast, crop canopies in agricultural systems frequently affect the local microclimate. In both cases, the soil has a significant impact on the climate close to the surface. The partitioning of incident radiation and how much energy is required to evaporate water, warm the air above the ground, or warm the soil are all influenced by the characteristics of the surface soil layer, such as color, water content, texture, and density. The soil heat flux, also known as heat flux density, is the volume of thermal energy that passes through a given area of soil in a given amount of time. How quickly the soil temperature fluctuates throughout the day or between seasons depends on how well it conducts heat. The rate of biological and chemical activities in the soil that are necessary for plant growth is significantly influenced by the soil temperature. Because it efficiently connects energy transfer mechanisms at the surface (surface energy balance) with energy transfer processes in the soil, soil heat flux is significant in micrometeorology (soil thermal regime). This interaction between methods for transferring energy from the surface to the subsurface has prompted in-depth studies of soil heat flux for a range of agricultural systems.

As opposed to being a genuine solid, soil is made up of individual particles and aggregates of organic and inorganic components that are divided by pores that are occupied by air and water. Depending on the shape of the soil voids and the total water content, the relative proportions of these final two components can vary greatly. Contrary to what might seem logical, the relative amount of pore space decreases as soil particle size increases. The rate of all biological, chemical, and physical soil processes necessary for plant development and the supply of ecosystem services is significantly influenced by temperature (McInnes, 2002). The speed of the soil temperature change throughout the day or with the seasons depends on the pace at which heat is transmitted in the soil. Because it efficiently combines energy transfer processes at the surface, the surface energy balance, with energy transfer processes in the soil, the soil thermal regime, soil heat flux, or the rate of thermal energy transfer across a unit area of soil over time, is a crucial metric. Therefore, the soil surface serves as a crucial contact for the passage of energy between the surface and the subsurface (Sauer and Norman, 1995; Sauer et al., 1995). The consequences of global climate change on temperature and precipitation patterns have a substantial influence on the conditions of the surface soil, which in turn have important repercussions for crop development and food security. Many economically significant crops and the ideal temperature ranges for growth are now well understood (Sánchez et al., 2014; Yamori et al., 2014). As indicated by the recent planting of commodity crops in unconventional production locations, such as increasing maize (*Zea mays* L.) and soybean (*Glycine max* L.) production in the northern Great Plains of the United States, growers and plant breeders are aware of these temperature ranges. Biological stressors or changing soil water conditions frequently occur in direct or indirect correlation with changes in mean and severe temperatures (weed competition, insects, and pathogens). Abiotic (temperature and moisture stress) and biotic (weeds, insects, and disease) stress interact in a complex way, which can make it difficult to develop effective control and mitigation measures through plant breeding or crop management techniques (Suzuki et al., 2014; Ramegowda and Senthil-Kumar, 2015). Any approach for adapting to climate change must take into account the thermal regime of the soil, how it interacts with the water accessible to plants, and the strategies available to control the soil's temperature and water content for the best crop yield (Kang et al., 2009). The crucial function of soil in agricultural systems within the context of the "water, food, and energy nexus" is still a topic of discussion on a global scale (Hatfield et al., 2017).

SURFACE ENERGY BALANCE AND SOIL HEAT FLUX

In micrometeorology, the surface energy balance is frequently taken into account while measuring soil heat flow;

$$R_n - G = LE + H$$

where R_n is the net radiation, G is the soil heat flux density at the soil surface, andLE and H are the latent and sensible heat flux densities, respectively. All terms have units of $Jm^{-2}s^{-1}$ or Wm^{-2}, also, all fluxes away from the soil surface are defined as positive except for R_n. The left side of Eq. [1], (Rn –G), represents the available energy while the terms on the right side (LE and H) are often referred to as the turbulent fluxes. Through terrestrial longwave radiation, a large portion of the energy that enters the soil during the day returns to the atmosphere at night. The short-term (i.e., hourly) energy balance calculations can contain significant errors if G is not taken into account because it is frequently the smallest component of the daily surface energy balance and has, in some cases, been disregarded. However, there are frequently significant transfers of energy into and out of a soil sample during both day and nighttime hours.

Since the 1950s, extensive investigations on the surface energy balance have been carried out. Early examples of these studies were Lettau and Davidson (1957) and Lemon (1963). Surface energy balance measurements have increased significantly in frequency in recent years due to technological advancements in ground-based and remote sensing instrumentation. With the advancement of sensor technology, the spatial scale of energy balance investigations has also increased. For instance, various attempts to estimate the earth's yearly mean energy budget have been made in response to interest in global climate change (e.g., Ohmura & Gilgen, 1993; Kiehl& Trenberth, 1997). The creation of the Global Energy Balance Archive (GEBA) to record present climatic circumstances and facilitate the research of historical and future climatic situations is proof of this tendency (Gilgen & Ohmura, 1999). For a variety of agricultural land use techniques, measurements of G, either separately or as part of energy balancing studies, can be found as examples. Data on G have been reported for woods (Tamai et al., 1998), orchards and vineyards (Heilman et al., 1994), grasslands (Twine et al., 2000), small grains (Kimball et al., 1999), row crops (Ham et al., 1991;), and lying sheep in various micrometeorological investigations (Gatenby, 1977). Studies on energy balance that have measured or estimated G for non-crop surfaces include those for bare soils, sloping terrain, and sparse canopies (Kustas et al., 2000), among others (Rao et al., 1977; Enz et al., 1988).

In addition to providing a wealth of information about soil heat flux under field settings, the quantity of research and the variety of soils and surface covers analyzed also testifies to the intricacy of the linkages in the energy balance. The complex interactions between the terms in the aforementioned equation, which are influenced by elements including soil and air temperature, soil water content, canopy properties, residue cover, and wind speed, can have a big impact on the fluxes' direction and magnitude. Surface cover, soil moisture content, and solar irradiance all affect how much G weighs in the surface energy balance. For a bare, dry soil in June, daytime peak hourly G values may exceed 300 Wm^{-2} (Fuchs &Hadas, 1972).

By contrast, hourly G for a moist soil beneath a plant canopy, residue layer, or snow cover will often be less than ±20 Wm^{-2}. Surface soil heat flux typically represents 1 to 10% of R_n for growing crops (Clothier et al., 1986). In the fall and spring when Rn is low and the soil is cooling or warming, or in desert climates where there is no vegetation, this proportion can go above 50%. (Choudhury et al., 1987).

The energy balance terms for a soil covered in corn (*Zea mays* L.) residue in central Iowa are shown in the figure below (Sauer et al., 1998). The information in the graph was collected in November, when the soil and fresh residue layer were both dry and the average surface temperature was 6.8°C. Daytime G averaged14.4% and 16.8%of R_n for Days 310 and 311, respectively.When it comes to research on water balance and water management in agricultural systems, measurement or prediction of evaporation is frequently of great interest.

Numerous evaporation measurement and prediction approaches require the input of soil heat flux, which is a part of the available energy. For example, evaporation measurements based on the Bowen ratio energy balance approach (Bowen, 1926) depend on a precise estimate of the energy that is available (R_n-G). Malek (1993) and de Silans et al. address how mistakes in G affect turbulent fluxes calculated using the Bowen ratio approach (1997). The Penman-Monteith (Penman, 1948; Monteith, 1965) and Priestley-Taylor (Priestley & Taylor, 1972) equations, two of the most popular equations for estimating evaporation, also require available energy as an input.

The impact of leaving out G will vary depending on the local climate, soil characteristics, and cropping system. Excluding G during the summer can result in systematic overestimation of available energy and subsequent overestimation of evaporation, even if failure to incorporate soil heat flux may create relatively minor (i.e., 10 percent) inaccuracies in available energy (Kumar & Rao; 1984; Anadranistakis et al., 1997).

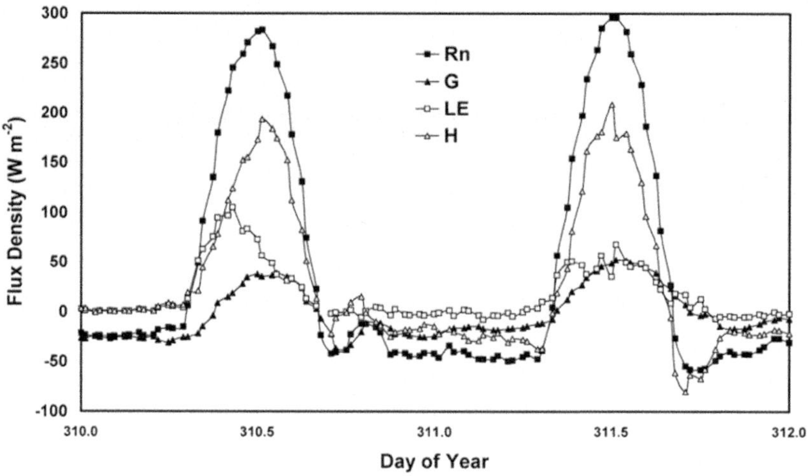

Figure 26: *Diurnal patterns of energy balance terms for a no-till corn field in central Iowa*

Source: Sauer et al., (1998)

SOIL THERMAL REGIME AND SOIL HEAT FLUX

Although radiation and convection can sometimes transmit energy in very shallow layers, conduction is typically the main method of energy transport in soils. When Fourier's Law is used, heat flow in soil can be compared to heat flow in a solid;

$$G = -\lambda \partial T / \partial z$$

where λ is the thermal conductivity of the soil ($Wm^{-1}K^{-1}$) and $\partial T/\partial z$ is the vertical temperature gradient (Km^{-1}) of the soil layer. In steady-state situations, Fourier's Law is specified for solid, homogenous materials with thermal conductivities that are basically constant throughout narrow temperature ranges. In many engineering applications, Fourier's Law may be used directly and quickly; however, using the aforementioned equation in a porous, three-phase media like soil is far more challenging.

The mineral type, particle size, and volume of organic matter all affect the solid component of soil's thermal conductivity, along with water content and bulk density (Abu-Hamdeh& Reeder, 2000). These characteristics frequently differ between different types of soils, geographically at the soil surface for the same soil, between soil layers, and over time. For instance,

Al Nakshabandi and Kohnke (1965) found that a silt loam soil's heat conductivity increased eight times as its gravimetric water content rose from oven dry to 0.34 kg kg^{-1}. Additionally, a temperature gradient in the soil causes water to move through evaporation and condensation, which simultaneously transmit latent energy (Cahill &Parlange, 1998). In most field situations, solar radiation is the primary energy source that drives soil heat transfer. Soil thermal characteristics are dynamic, but also annual and diurnal variations of solar radiation, including erratic weather patterns, are superimposed on these complicated connections.

It is possible for agricultural management techniques like tillage, irrigation, and drainage to have an impact on the thermal characteristics of soils and, consequently, the thermal regime of the soil. Numerous researchers have focused on, in particular, how tillage and crop residue management affect soil heat flux (Azooz et al., 1997; Richard & Cellier, 1998; Sauer et al., 1998). The soil's top layer is loosened by tillage, while some localized compaction may also happen. Lower soil bulk density generally translates to lowerλ, thus, lower G has been observed in tilled soil as compared with un-tilled or compacted soil (Azooz et al., 1997; Richard & Cellier, 1998). Crop residue has a low thermal conductivity and may prevent heat transfer into the soil whether it is on the soil surface or has been incorporated via tillage. Additionally, residue layers are a barrier to vapor passage and have a shortwave reflectance that is higher than that of most soils (Sauer et al., 1997). As a result, soils with a lot of crop residue on the surface typically have greater water contents, cooler temperatures, and lower G. Naturally, such variations in the thermal regime of the soil have an impact on evaporation and the surface energy balance.

MEASUREMENT OF SOIL TEMPERATURE AND THERMAL PROPERTIES

Thermocouples are most frequently used to measure soil temperature; however, thermistors or resistance temperature detectors may also be used (RTDs). Thermocouples are created by joining together wires made of two dissimilar metals, frequently one or more of which is an alloy, to create a closed circuit. This creates an electromotive force (emf) that is specific to the metal pair (couple) and inversely proportional to the temperature difference between the two wire junctions. This See-beck voltage (mV or μV $°C^{-1}$) can be accurately monitored with modern dataloggers and converted to temperature using an independent reference temperature provided by the datalogger. The accuracy of thermocouple temperature measurements is ~ 0.75% of the measured temperature. For resistors or metals like platinum, copper, and nickel, thermometers and RTDs use variations in electrical resistance with temperature (McInnes, 2002). The resistance change of the sensor elements is measured using bridge circuits. Before being introduced into the soil, thermocouples, thermistors, and RTDs must all be covered in a waterproof substance. The sensors and the wires that connect them to a datalogger are typically not harmed during the process of backfilling a trench that has been excavated to the necessary measurement depth. Thermocouples are accurate enough for the majority of applications, inexpensive, and durable with an acceptable level of failure under field circumstances, while not being as sensitive as thermistors or RTDs. When higher accuracy is desired, thermometers or RTDs are preferred since they are more expensive, more stable, but less robust (0.5 percent of measured temperature). These are applications that measure temperature change over brief time periods and/or modest depth increments, particularly close to the soil surface (Kool et al., 2016).

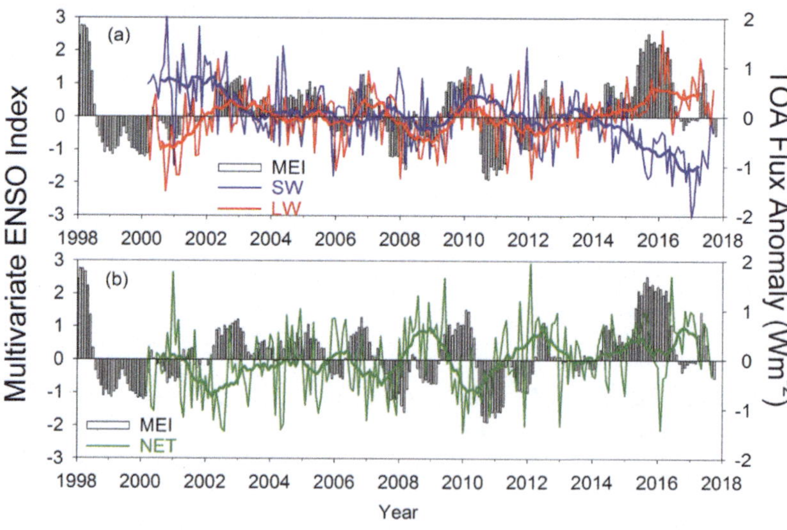

Figure 27: *Global mean* ***(a)*** *shortwave (SW) and longwave (LW) and* ***(b)*** *net top-of-atmosphere (TOA) flux anomalies for March 2000–September 2017 from CERES Energy Balanced and Filled (EBAF) Ed4.0. Thin lines denote monthly anomalies, thick lines are 12-month running means. Vertical black bars show the Multivariate ENSO Index (MEI). Anomalies are calculated relative to climatology over the entire period. SW and LW TOA flux anomalies are defined as positive upwards and net TOA flux anomalies are positive downwards.*

Source: Loeb NG, Thorsen TJ, Norris JR, Wang H, Su W. Changes in Earth's Energy Budget during and after the "Pause" in Global Warming: An Observational Perspective. Climate. 2018; 6(3):62. https://doi.org/10.3390/cli6030062

Soil thermal conductivity (l_s) is a physical property relating to how quickly the soil can conduct thermal energy (units of $Wm^{-1}K^{-1}$) while the heat capacity of soil (C_s) is defined as the amount of energy required to raise a unit volume of soil by one degree Kelvin (units of MJ m^{-3} K^{-1}). Values of l_s vary significantly with soil water content owing to the large difference in λ between water and air (0.57 vs.0.025 Wm^{-1} K^{-1} at 283K, respectively). Water content variations can cause ls values in the field to vary by a factor of 2 to 5. Such significant variations can happen quickly at shallow levels when the soil's water content changes as a result of precipitation infiltration, irrigation water use, drying due to evaporation, and plant water intake (Abu-Hamdeh and Reeder, 2000; Abu-Hamdeh, 2003). Cs is also strongly influenced by soil water content as C values for the solid, water, and air fractions of soil are (2.1 to 2.5 MJ m^{-3} K^{-1}, 4.2 MJ m^{-3} K^{-1}, and 0.0012 MJ m^{-3} $K^{-1)}$, respectively. de Vries (1963)provided a widely-employed technique for estimating C_s based on the volume fractions of the mineral constituents and water and their respective heat capacities

$$C_s = 2.35\,(1 - f) + 4.18\theta$$

where f is the porosity and ϴ is the volumetric water content (m^3 m^{-3}). Measurements of soil thermal properties in the laboratory and field are now commonly accomplished using the dual-probe heat-pulse technique (Bristow et al., 1994; Ochsner et al., 2001; Bristow et al., 2001; Smits et al., 2010). This technique is based on a solution of de Vries (1952) for the change in temperature as a function of time at a radial distance from the heat pulse source;

$$T(r,t) = \frac{q}{4\pi\alpha C_s}\left[Ei\left(\frac{-r^2}{4\alpha(t - t_0)}\right) - Ei\left(\frac{-r^2}{4\alpha t}\right)\right] \quad t > t_0$$

where r is the radial distance (m), t is time (s), t_0 is the duration of the heat pulse (s),q is the quantity of energy applied (W m^{-1}), α is the soil thermal diffusivity equal to l_s/C_s (m^2 s^{-1}), and Ei(-x) is the exponential integral (Ochsner et al., 2001). Welch et al. (1996) presented a nonlinear regression technique to fit an analytical solution of the above equation to the temperature increase versus time data from the sensor that produces estimates of C_s, α, and l_s. This general approach is now widely used with variations in sensor design and data analysis techniques.

Soil Heat Flux Measurement Techniques

Four techniques have been used to measure the majority of soil heat flux densities (flux plate, calorimetric, gradient, or combination). The block method is an additional technique that has been created but is not commonly used (van Wijk, 1967). The measurement of soil temperature, one or more thermal parameters (such as thermal conductivity or heat capacity), and maybe soil water content are required to determine the soil heat flux density. The spatial variability of G under field settings is a recent subject of study in soil heat flux measurement. When LE and H are measured using eddy covariance or Bowen ratio techniques, which typically represent an area of 100s of m^2, concern is increased. In contrast, a single measurement of G represents a much smaller area, possibly 0.1 m^2, whereas R_n (measured with a net radiometer) may be representative of an area of 10s of m^2. Several investigations have demonstrated that regional variation of G under field settings can be significant (Ham &Kluitenberg, 1993; Tuzet et al., 1997; Kustas et al., 2000). This is especially true at locations with sparse canopies or uneven terrain, where considerable variations in soil temperature and water content are introduced by both shade patterns and microtopography. In a dune with an uneven surface and partial shrub cover, Kustas et al. (2000) observed that variation in G (measured at 0.08 m depth) between neighboring places with similar cover was larger than 200 W m^{-2}. In order to get a spatially-representative value of soil heat flux during energy balance studies for agricultural surfaces, especially partial canopies, it is important to measure soil heat flux density at many places in light of the evidence supporting the geographical variability of G.

Staley and Gerhardt (1957), Carson and Moses (1963), Kimball and Jackson (1979), Fuchs (1986), and Sauer (1986) are a few authors that have previously reviewed the methods for measuring soil heat flow (2002). Each of these reviews concentrated on the measurement of G as it relates to the characterization of soil thermal regime, with the exception of Staley and Gerhardt (1957) and Carson and Moses (1963). Although the surface energy balance and soil thermal regime are interdependent, the emphasis in this section will be on measuring G in relation to micrometeorological applications (surface energy balance and evaporation).

Flux Plate Method

The flux plate method is the most popular G measurement method (Sauer, 2002; Sauer and Horton, 2005). The idea of a soil heat flux plate was modified from research into measuring heat transfer in ship bulkheads and building walls. The first person to use this method precisely to measure heat transport in soil is credited to Falckenberg (1930). Dunkle (1940), Deacon (1950), Portman (1958), Philip (1961), Fuchs and Tanner (1968), and Mogensen have all contributed improvements to the theory and design of soil heat flux plates (1970). Soil heat flux plates are tiny disc-shaped sensors that are buried horizontally in the soil at a depth of 1 to 10 cm. They are also known as heat flow meters or heat flow transducers. The fact that the plates are convenient and reasonably priced makes this method popular. Most flux plates have a thermopile, which is a collection of thermocouples wired in series to provide an emf proportional to the temperature difference across the plate. The heat flow through a calibrated plate of known and fixed can then be measured, and G for the surrounding soil can be inferred, using the voltage signal from the plate. In spite of the simplicity and wide acceptance of the heat flux plate method, several areas of concern surround this approach:

- heat flow convergence/divergence around the plate,
- water flow divergence (including water vapor),
- thermal contact between the plate and soil matrix, and
- accounting for heat storage in the layer(s) between the plate and the soil surface.

Concerns about convergence/divergence of heat flow are raised by the proximity of an impermeable plate to the soil surface, which could distort estimations of G derived from plate readings. Based on a theoretical examination of heat flow near a tiny oblate spheroid with known thermal properties, Philip (1961) suggested a number of elements to take into account when designing plates to minimize disruptions in vertical heat flow. The ratio of heat transfer through the plate to that in the soil was predicted using the following equation:

$$\frac{G_m}{G} = \frac{\varepsilon}{[1 + (\varepsilon - 1)]H}$$

where G_m is the heat flux density through the plate, ε is the ratio of the plate thermal conductivity to the soil thermal conductivity (λ_m/λ), and H is an empirical factor related to plate shape:

$$H = 1 - (\beta b)$$

where β is a dimensionless geometric constant dependent on plate shape and b is the plate thickness divided by side length for a square plate or plate thickness divided by plate diameter for a circular plate. Mogensen (1970) tested (G_m/G) in laboratory experiments with a small circular plate and found that the heterogeneous composition of the plate made it difficult to verify Philip's theory. Watts et al. (1990) confirmed and refined Philip's recommendations to conclude that:

- the plate thickness to width ratio should be small,
- dry and saturated sand are suitable calibration media for mineral soils, and
- the thermal conductivity of the plate should be greater than 0.5 W m^{-1} K^{-1}.

The relationship between thermal conductivity and the ratio of heat flow through the plate to that through the soil (Gm/G), as predicted by the equation, is shown in the figure below. The curves in the figure below are for hypothetical circular plates that are 50 mm in diameter, 2, 4, or 6 mm-thick, and have thermal conductivity (λm) of 0.25 or 0.75 $Wm^{-1}K^{-1}$. As the difference between the plate and soil λ increases, increasingly large errors (up to –50%) in the measured Gare predicted. Clearly, using a heat fluxplate with λ similar to the expected soil λ or within a range common for mineral soils (~0.4–1.2 $Wm^{-1}K^{-1}$) would reduce errors associated with heat flow convergence/divergence through the plate. The curves in the figure below also indicate that there is less error associated with thinner plates. For example, if soil had a λ of 1.0 W m^{-1} K^{-1}, a 2 mm-diameter plate with a λ_m of 0.25 $Wm^{-1}K^{-1}$ would have22% less error in measured G than a 6mm-diameter plate with the same λ_m.

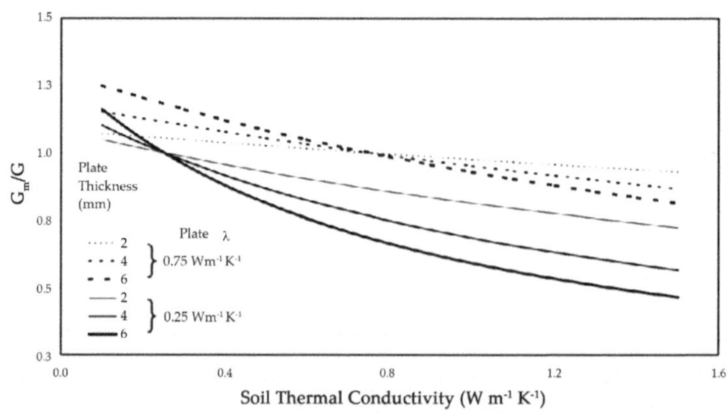

***Figure 28:** Effect of heat flux plate thickness and thermal conductivity on the ratio of the heat flux*

Source: Sauer, & Horton, (2005).

It should be noted that Philip's analysis assumes that the flux measurement is made across the entire plate area, while several current plate designs include a "guard" area surrounding a thermopile located in the center of the plate. Also, this analysis is only for heat conduction and does not include any provision for the effects of energy transfer due to liquid water movement near a plate.

The presence of an impermeable object, such as a heat flux plate, close to the soil surface can have a variety of effects on how water flows through the soil. A heat flux plate can prevent water from moving downward in the soil during and after precipitation events, which causes the soil directly above the plate to have more water than the soil immediately below. The plate can restrict upward water movement as evaporation dries the soil, causing the soil above the plate to become drier than the surrounding soil while the soil below the plate stays moist. It will also have varied thermal properties if the soil directly around the plate has a different water content than the soil around it. For a typically loam soil, a change in water content from 0.15 to 0.20 kg^{-1} can increase itsλ from 0.96 to 1.3W $m^{-1}K^{-1}$ and its volumetric heat capacity (C) from 1.75 to1.96 MJm^{-3} K^{-1} (de Vries, 1963; AlNakshabandi & Kohnke, 1965). Unrepresentative soil water content next to a heat flux plate will have an impact on the thermal properties of the soil and may cause significant inaccuracies in the observed G because it will modify the heat flow in the area around the plate. The latent heat of vaporization (also known as evaporative cooling) occurs in the soil underneath a soil heat flux plate if it is positioned above the drying front. The calorimetric heat storage adjustment, which is only performed for the soil layer above the plate, does not account for this energy loss either because the source of the energy may be located deeper in the soil. Large errors in G (up to 100 W m^{-2}) can occur when latent heat loss from bare soil during periods of high evaporation is disregarded, as demonstrated by De Vries and Philip (1986) and Mayocchi and Bristow (1995). However, such large errors are only likely when the plate is positioned close to the surface (0.02 m), evaporation is high, and the drying front is abrupt. The inaccuracy from latent heat loss shouldn't be considerable if there is a plant canopy, a diffuse drying front, or if the flux plate is at least 0.05 m below the surface.

The soil's texture, structure, water content, and the plate's installation technique all affect the thermal contact between the plate and the soil. Comparing laboratory and field calibrations of soil heat flux plates, Fuchs and Hadas (1973) came to the conclusion that contact resistance was the primary cause of error in the field observations. The calibration media should be chosen so that the thermal contact is similar to that of the soil where the plate is to be mounted in order to prevent mistakes brought on by contact resistance. Alternately, plate calibrations might be finished on-site by contrasting the G values obtained using flux plates and those obtained using one of the other techniques. The calorimetric method has been employed to account for the change in heat storage in the soil above the heat flux plate because the heat flux plate only measures G at the depth of installation (Fuchs & Tanner, 1968). Large inaccuracies can occur if the soil above the plate's heat storage is not taken into account. When the change in heat storage was disregarded, Mayocchi and Bristow (1995) found inaccuracies as big as 80 W m-2 in daytime, half-hour average G in an Australian sugar cane (*Saccharum officinarum* L.) field. When assuming that the change in temperature of the soil layer above the plate is well-approximated by the average temperature at the midpoint, Massman (1992, 1993) found that the conventional calorimetric adjustment may have inaccuracies of 3 to 10 percent. An analytical approach for calculating G at the surface is presented by de Silans et al. (1997) using G measured by a plate, temperature at the soil surface, and plate depth. This method doesn't require any understanding of the thermal characteristics of soil, but it does call for harmonic analysis of waves of heat flux and constant, periodic temperature over several days.

Sensors

Deacon (1950), Tanner (1963), and Fuchs and Hadas (1973) explored the design and construction of soil heat flux plates. In general, plate design is dictated by the requirements that the plate be thin, watertight, have a λ_m comparable to the soil being monitored, and that the emf produced is high enough to be easily measured. To assure good thermal contact with the soil, exterior materials with a high thermal admittance are desirable. Thermal admittance (A_T) is defined as;

$$A_T = (\lambda_m C_m)^{0.5}$$

where C_m is the heat capacity of the plate (Fuchs, 1986). To obtain a high A_T, several plate designs have used metal and/or anodized metal shields on their exteriors. Numerous sensors with various dimensions, thermal characteristics, and sensitivities are now offered commercially as a result of the success of the flux plate approach. The soil thermal and water regimes being observed, as well as the intended depth of installation, are factors to take into account when choosing a soil heat flux plate. Plates with larger areas and lower λ_m positioned at 0.025 to 0.05 m depth may be suitable for G measurements with dry soils in arid environments as the soil λ and latent heat loss are likely to be relatively low. Smaller plates with higher λ_m between 0.05 and 0.1 m may be preferable at a humid site where more frequent rainfall events keep the soil moist so the soil λ is higher and the drying front is above the plate depth.

Procedure

One configuration of sensors for use with the heat flux plate approach is shown in the image below. A horizontal slit in a side wall that is slightly smaller than the plate dimensions is produced after a shallow excavation that is made to a depth below the desired depth of plate placement. To ensure that there is good thermal contact with the soil on all sides and that the plate faces are parallel to the soil surface, the plate should be gently put into the slit. The placing depth for soil heat flux plates is normally between 0.025 and 0.1 meters. There is worry that the soil at depths below 0.02 m may break when dry, exposing the plate to sunlight or resulting in inadequate thermal contact between the plate and soil. To accurately reflect the location being monitored, many plates in the same excavation or multiple single-plate excavations could be required. The distance between neighboring plates should be at least twice as great as the largest plate face size (diameter or side length) (Watts et al., 1990).

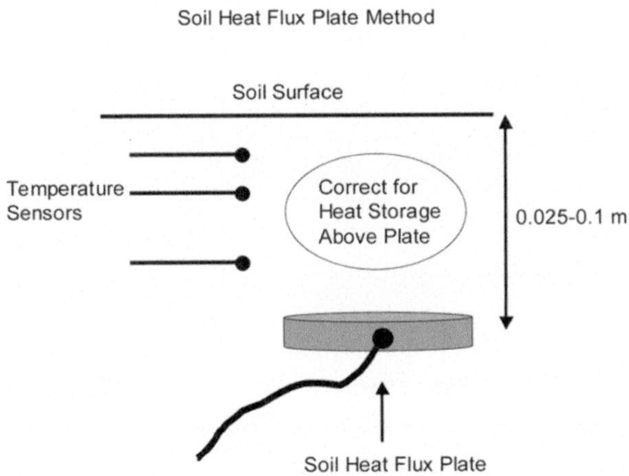

***Figure 29:** Arrangement of sensors for application of the heat flux plate method.*

Source: Sauer, T. J., & Horton, R. (2005).

The distribution of measurement locations should account for any spatial heterogeneity brought on by microtopography or shading. To lessen the chance of heat conduction along the wire to or from the sensor, a section of sensor wire next to the plate should be buried at the same depth as the plate. To assist the calorimetric heat storage correction, soil temperature is frequently recorded with thermocouples or thermistors above the heat flux plates (Fuchs & Tanner, 1968). Generally, depending on the depth of the plate and the consistency of the soil qualities, one to four measurement depths are utilized. The temperature sensor profile should ideally be placed next to each plate, but depending on the distance between the plates and the spatial variability of the soil parameters, more than one plate could have their heat storage corrected using the same temperature profile. McInnes (2002) gives thorough information on how to assess soil temperature.

Calibrating Soil Heat Flux Plate

The gradient and calorimetric approaches (Robin et al., 1997) have been used to calibrate soil heat flux plates in relation to G. (Kimball et al., 1976; van Loon et al., 1998). Any inaccuracies related to thermal contact resistance should be minimized by an in-situ calibration (Fuchs &Hadas, 1973). The best way to collect reliable calibration data at different soil moisture contents is also through such calibrations; but, in-situ calibrations require a significant time and resource commitment, and measurement errors related to the method employed as the standard must be taken into account. A heat flux plate with a heater for independent in-place calibrations is a recent advancement in sensor technology (van Loon et al., 1998). Other methods for calibrating soil flux plates have been put forth, such as a radiation technique (Idso, 1972) and other steady-state laboratory apparatus designs (van Loon et al., 1998). The plates are placed in a porous medium (often quartz sand) using laboratory techniques inside a soundly insulated box where a known, 1-dimensional heat flow is maintained across the media and through the plates. Accurate calibrations using this technique can only be carried out under dry and saturated conditions because the formation of a temperature gradient across moist sand produces a moisture gradient and latent heat transfer.

Advantages and Disadvantages

The fact that the plates are reasonably affordable, dependable, and may be used continually in the field for lengthy periods of time are the main benefits of the flux plate method. Due to these factors, the flux plate approach has emerged as the most often used method for determining G, particularly in applications involving surface energy balance. However, unless specific safety measures are taken and/or corrections are done, there are likely to be considerable mistakes in measured G:

- The thermal characteristics of the plate should match the measured ambient conditions. When estimating errors using plates whose heat conductivity is significantly different from the soil around them, errors as large as 50% are calculated. When deciding on the placement depth, plate thickness, face area, and near-surface soil water dynamics should all be taken into account.
- It is important to accurately calibrate and install plates. A flux plate's sensitivity varied by 27 percent according to Fuchs and Hadas (1973), who calibrated it both in the field and in a controlled laboratory setting. Differences in thermal contact resistance with the soil were found to be primarily responsible for this inaccuracy.
- Installing enough plates will enable you to get a geographically accurate estimate of G. Spatial variation of G can be large, especially with partial canopies and uneven soil surfaces, demanding the careful positioning of many sensors to produce a correct areal average.
- Correction is required for heat storage above the plate. The heat storage adjustment results in much higher G during the day and lower G at night, despite the fact that the daily sum of G for the corrected flux is only 3.3 percent greater than the uncorrected flux. If the turbulent fluxes were calculated using the Bowen ratio energy balance approach, failing to account for heat storage would result in an overestimate of the available energy of 52 W m^{-2} (14%) (R_n - G = 377 W m^{-2} instead of 325 W m^{-2}). This inaccuracy would result in a subsequent equal-sized underestimation of LE + H.
- For shallow plates above an abrupt drying front, latent heat loss may need to be corrected. If latent heat transport is overlooked in these circumstances, inaccuracies as large as 100 Wm^{-2} can arise, as demonstrated by Buchan (1989) and Mayocchi and Bristow (1995). When plates are positioned below 0.05 m, this inaccuracy is probably negligible.

Calorimetric Method

From the change in heat storage in the soil during a certain time period, the calorimetric or temperature integral approach is used to calculate the average soil heat flow density (Fuchs, 1986):

$$G_{i-1} = \delta z_i C_i (\partial T_i / \partial t) + G_i$$

where G_{i-1} is the heat flux density at the top of a layer, δz_i is the layer thickness(m), C_i is the volumetric heat capacity for the layer (J m^{-3} K^{-1}), $\partial T_i / \partial t$ is the rate of change of the mean layer temperature (K), and G_i is the heat flux density at the bottom of the layer. If a layer n is found with $\partial T/\partial z = 0$, then by G, $G_n = 0$ and the soil heat flux density for layer j is given by;

$$G_{j-1} = \sum_{i=j}^{n} \delta z_i C_i (\partial T_i / \partial t)$$

When j = 1, the above equation gives G at the soil surface. Alternatively, if G_i is known by some other method at a reference depth (z_r), (G_{i-1}) can be used to calculate G at all other depths. If G_r is calculated using the gradient method, this approach is referred to as the combination method.

Sensors

The only data inputs required to complete the calculations involved with the calorimetric analysis are soil temperature and volumetric heat capacity. For identifying diurnal trends in G, soil temperature must be precisely measured at many depths, often at or below 1 m, from close to the surface to a depth where G is negligible for the required measurement interval. For the surface layers, where temperature changes quickly, different temperature sensors may be preferred (e.g., more sensitive thermistors or smaller thermocouples) than for deeper in the profile, where the amplitude of soil temperature is dampened. Again, McInnes (2002) provides thorough information on how to monitor soil temperature. Kluitenberg discusses methods for calculating soil heat capacity (2002). In both lab and field settings, heat pulse probes or calorimetry are employed to measure the soil's heat capacity (Kluitenberg et al., 1993; Bristow et al., 1994). The method by De Vries (1963) for calculating volumetric heat capacity is based on the volume fractions of a soil's mineral, organic matter, and water components is also commonly employed. The soil porosity, organic matter content, and volumetric water content must all be known in order to perform the de Vries estimating technique. So, in order to determine a soil's volumetric heat capacity, only the measurement of the volumetric water content is required if the soil's bulk density and organic matter content are known. Several invasive or non-invasive approaches can be used to measure the amount of water in the soil (Topp& Ferré, 2002). To enhance spatial averaging, multiple sensors or water content samples obtained from each site and depth may be used.

Procedure

At the location(s) where sensors will be inserted into an exposed soil face, a reasonably deep excavation will be necessary. The presence or absence of a plant canopy or residue layer, surface roughness, row spacing and orientation, and proximity to anomalous features should all be taken into account when choosing measurement locations (e.g., large rocks, tree roots, and compacted areas). To successfully apply the approach, depth intervals must be carefully chosen. To guarantee equal volumetric heat capacity and linear temperature gradients within each layer, abrupt changes in soil particle size, mineralogy, or bulk density should be avoided. Near the surface, water content and temperature changes are typically more pronounced, which causes G per depth increment to decrease as one moves away from the surface. In a homogeneous soil stretch, the best sensor placement should resemble a geometric progression with depth (e.g., 0.01-, 0.02-, 0.04-, 0.08-, 0.16-, 0.32-, 0.64-, 0.96-, and 1.28-m depths). The time constant of the temperature and heat capacity sensors, as well as the protocol's goals, will determine the measurement frequency and averaging interval (e.g., hourly flux vs. daily sums vs. seasonal trends). Data signals are typically averaged over periods of 15 to 60 minutes. Identification of a layer with G = 0 or determination of G at a reference depth then permits calculation of G at all other depths and at the soil surface. If heat capacity is being estimated, soil water content and bulk density values must be available for each layer between the temperature sensors (Grossman &Reinsch, 2002).

Advantages and Disadvantages

The ability to precisely predict changes in volumetric heat capacity for each soil layer and each time interval limits the calorimetric approach. This may not be a major problem in situations where the soil's water content is slowly changing; however, it can be challenging to obtain accurate heat capacity values in layered soils or in situations where the soil's water content is changing quickly (such as when a wetting front advances after a rainfall event). A contemporary example of applying the calorimetric approach to estimate G is given by Cellier et al. (1996), who also developed an estimating method employing widely used micrometeorological factors. For this procedure to work, extremely accurate soil temperature readings (≈0.02-0.1 K accuracy) are needed. This is crucial at deeper depths because layers with high volumetric heat capacities are progressively made thicker as a result of the temperature difference. This is crucial if data are needed for intervals of time with minor variations in temperature (Hanks & Jacobs, 1971; Fuchs, 1986). The calorimetric method is most frequently used as a benchmark for comparison with other approaches due to the very strict standards for temperature readings, the high number of sensors, and their location deep in the soil.

Gradient Method

The gradient method is a direct application of the G equation above. A measured temperature gradient ($\partial T/\partial z$) is combined with an estimated or measured thermal conductivity (λ) to determine G. The simplicity of this approach is offset by difficulty in obtaining accurate λ measurements under field conditions. Like volumetric heat capacity, soil thermal conductivity changes significantly with soil water content, in this instance due to the large difference in λ between water and air (0.57 vs. 0.025 W m^{-1} K^{-1} at 283 K, respectively). As a result, λ values for soil layers in the field can change by a factor of two to five across commonly observed changes in water content.

Sensors

Details on measuring soil thermal conductivity using steady-state and transient techniques are provided by Bristow (2002). For in-field measurements, the single heat probe and heat pulse approaches work best. The approximate method by de Vries (1963) for calculating soil thermal conductivity based on a phase mixing model and the weighted volume fraction of soil elements is also covered in the discussion. Data on soil water content will be required if the de Vries method is to be used for estimating λ. Thermocouples or thermistors, which have adequate accuracy and are simple to log for continuous data, are once again excellent temperature sensors for the gradient approach.

Procedure

The gradient method is often used at a reference depth of at least 0.2 m because it might be challenging to measure precisely close to the soil surface. G at the surface can be calculated using the calorimetric method on the surface layers above the reference layer. The combination method, which combines the calorimetric and gradient approaches, is covered in the following section. Depending on the soil's texture, moisture content, and temperature wave amplitude, a shallower or deeper reference depth may be suitable or even feasible. Layering within the soil and the overall goals of the G measurements will determine where to place the soil temperature, thermal conductivity, and/or soil water content sensors. To implant sensors at the correct depths and in strategic positions that accurately depict the measuring area, an excavation must be made. The spatial sampling will be improved by placing additional sensors on the same horizontal plane or in additional locations. Based on sensor specifications, sampling goals, and the rate at which soil moisture and temperature fluctuate over time, sampling frequency and average intervals will be determined. The average temperature difference between each layer over the averaging interval or a smooth curve fit to the temperature data can both be used to determine the temperature gradient.

Advantages and Disadvantages

The gradient method is simple to employ, however, accurate measurement of λ in situ can be challenging. Recent improvements in thermal conductivity measurement techniques, especially for in-situ measurements, now make the application of the gradient method more attractive. Nonetheless, accurate estimates or measurements of λ near the soil surface are necessary if G is to be determined at the soil surface for use in energy balance or evaporation equations. Kimball et al.(1976) used four variations of the gradient method to measure G in a bare loam soil. They concluded that the method produced acceptable results with λ values estimated by de Vries' theory when a 0.2 m reference depth was used but not with a 0.05 m reference depth.

Combination Method

The combination method is the union of the gradient and calorimetric approaches. The combination method has been used to refer to the coupling of the soil heat flux plate and calorimetric methods (Fuchs & Tanner, 1968; Massman, 1993). The former will be referred to as the combination method in this instance, whereas the latter will be seen as a crucial heat storage adjustment for the proper use of the flux plate approach. The gradient and calorimetric methods are combined to benefit from the gradient method's few measurements and the calorimetric method's near-surface G measurements' accuracy. The calorimetric approach is used to measure G at successive layers between the reference depth and the soil surface, while the gradient method is used to find G at some reference depth (usually 0.2 m) through the application of the G equation. The combination approach enables the determination of G at the soil surface based on a flux obtained from the gradient method applied to one layer, avoiding the deep profile measurements of soil temperature and volumetric heat capacity of the calorimetric method.

Sensors

For the calorimetric and gradient methods, suitable sensors/techniques for measuring soil temperature, thermal conductivity, and heat capacity can be utilized with the combination method or its null-alignment variant (Kimball & Jackson, 1975).

Procedure

Due to the fact that all relevant sensors will be at or below 0.25 m, just a small amount of excavation will be required. Once more, the variety of soil qualities and surface cover within the area to be represented by the data will determine the choice of location(s) and number of sensors at each depth. The soil qualities will decide the number and vertical spacing of temperature sensors above the 0.2 m reference, but a geometric progression from the surface is still advised with a maximum vertical separation of 0.05 m. Sensors positioned above and below the 0.2 m reference depth can measure the temperature gradient there (e.g., 0.15 and 0.25 m). In order to determine the volumetric heat capacity for the calorimetric calculations, the soil water content and bulk density must be measured or estimated for each layer at the same time as the temperature readings, unless the heat capacity is directly measured.

Advantages and Disadvantages

Because the calculations for heat storage are applied to relatively thin strata close to the surface, measurements of soil temperature with a resolution of 0.1 K are appropriate. Success with the combination technique will depend on how precisely the thermal conductivity and volumetric heat capacity of the soil layers can be measured or estimated, as the required soil temperature measurements are simple. The null alignment method was compared with the theory of Philip and de Vries (1957) by Pikul and Allmaras (1984) and de Vries and Philip (1986), with contradictory results. In particular, for shallow soil layers and dry soil conditions, Pikul and Allmaras (1984) observed poor agreement between the null-alignment and mechanistic techniques. De Vries and Philip (1986) came to the conclusion that the null alignment technique's systematic mistakes and substantial underestimate were caused by the failure to account for latent heat loss in the higher soil layers.

Estimation and Prediction of G

Due to the importance of G to surface energy balance and evaporation investigations, numerous prediction techniques have been created for use in circumstances where observed values are not available. These techniques rely on fabricated micrometeorological data, remote sensing parameters, or soil thermal properties. One of the simplest approaches involves the ratio of G to R_n. This ratio is an indicator of the relative proportion of available energy that is used to warm the soil. During spring, G/R_n for the crop surface changes as the soil transitions from wet and cold with no plant canopy to warm and moist with increasing shade provided by the growing crop. Soil heat flux typically represents 1 to 10% of R_n for growing crops (Sauer and Horton, 2005). G/R_n can exceed 50% in the fall or spring when R_n is low and the soil is cooling/warming or in arid climates when there is little vegetative cover. Three soil heat flux plates (HFT-1, Radiation and Energy Balance Systems, Seattle, WA) were positioned in the crop row at 0.05 m depth to detect soil heat flow during a tillage trial in Minnesota (Sauer and Eash, unpublished data, 2001). Net radiation was measured with a net radiometer (Q*6, Radiation and EnergyBalance Systems, Seattle, WA) installed 1.5 m above the soil surface. For the 2000 growing season, significant differences in G_0/R_n were measured between each tillage system with mean values for the entire measurement interval of 0.23, 0.20, and 0.17 for no till, chisel, and strip till, respectively. The 2000G0/Rn values, however, showed no trend with time after planting of the corn.

Note that values in the figure below are only for daylight hours ($R_n >$ +100 W m^{-2}) on days without rain. In contrast to the 2000 data, in 2001 there were no significant differences between tillage systems, but a clear decrease inG0/Rn with time was observed. Mean values of G_0/R_n for the strip till and chisel tillage systems decreased from 0.21 to 0.10 from 2 to 56 d after planting. The significantly higher G_0/R_n for no till in2000 was due to lower Rn observed for the no till system (daily average 9.47 MJ m^{-2} compared to 9.79 and 9.90 for strip till and chisel, respectively). The no-till soil was the coolest of the three tillage systems, which would reduce the terrestrial longwave radiation and therefore reduce R_n. This system also had the greatest residue cover (78%), which would increase the amount of reflected shortwave radiation and also reduce R_n.

These relatively small differences along with slightly higher daily average G_0 (0.68 MJm^{-2} vs. 0.45 and 0.66 for strip till and chisel, respectively) combined to produce the significantly greater G0/Rn in 2000. Although the average daily values of G_0/R_n for all tillage systems in 2000 are comparable to those observed in 2001, it is not clear why the ratio failed to decrease with canopy growth as was observed in 2001. Sauer and Horton (2005) provide a summary table of example values of G/R_n for a variety of agricultural surfaces. G/R_n is relatively high (> ~0.15) for bare soils and sparse canopies and lower for fully-developed crop canopies and forests, which have greater attenuation of Rn within the canopy.

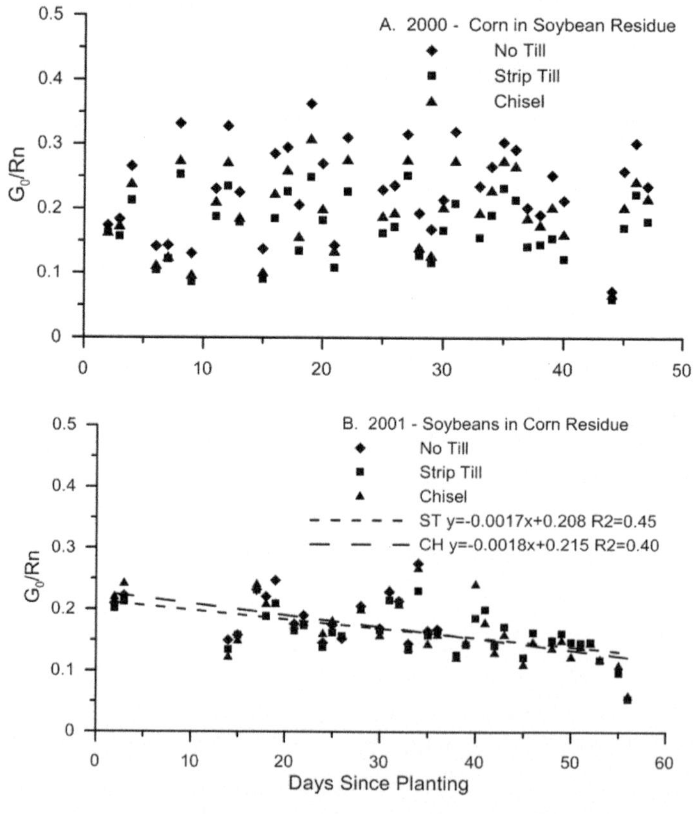

Figure 30: *Daily average G_0/R_n for three tillage systems.*

The ratio has been found to be sensitive to changing soil water content and canopy density (Clothier et al., 1986; Oliver et al., 1987), thus using G/Rn ratios to estimate G is primarily useful only as a first approximation. Daughtry et al. (1990) and Kustas and Daughtry (1990) extended the work of Clothier et al. (1986) to use multispectral reflectance data to determine an index based on the difference between the near-infrared and red reflectance divided by their sum. These two studies are two examples of more complex methods for estimating G.

The use of the normalized difference vegetation index (NDVI) resulted in improved estimates of G from Rn for fields with bare soil and crop canopies at different growth stages. Kustas et al. (1993) further explored nonlinear relationships between G/R_n and vegetation indices (VIs) and concluded that a power function was more appropriate than the previously derived linear relationships between G/R_n ratios;

$$\frac{G}{R_n} = aVI^b$$

where a and b are fitted constants.

In order to estimate G, Horton and Wierenga (1983) developed a harmonic analysis of the soil temperature at one depth or at two depths if an is unknown. Using hourly soil temperature data at three depths and a finite-difference solution to the transient heat flow equation, Sharratt et al. (1992) calculated G. By using fractional calculus, Wang and Bras (1999) established an explicit link between surface temperature and soil heat flux. Although they assumed that the soil thermal characteristics were consistent and independent of soil water content and temperature, the G estimated with their technique compared favorably with field data from Kansas, the United States, and from Brazil.

Soil Temperature and Heat Flux In Arable Cropping Systems

Irrigation, drainage, tillage, and residue management are among agricultural management techniques that have a significant impact on the soil thermal regime (Bristow, 1988; Unger, 1988). Lower ls are produced by tillage because it loosens the top soil layers and reduces their bulk density. Comparing tilled soil to untilled and uncompacted soil, it has been found that tilled soil has less G. (Sauer et al., 1996; Azooz et al., 1997; Richard and Cellier, 1998). Dry agricultural residue has a low l, and when present in significant amounts, whether on the soil surface or integrated into the soil by tillage, it can block heat transfer. Additionally, crop residue on the soil's surface has a higher shortwave reflectivity than most soils, which acts as a barrier to the movement of water vapor (Gausman et al., 1975; Horton et al., 1996; Sauer et al., 1997). Higher water content, lower temperatures, and lower G seem to be characteristics of soils with heavy residue cover. The surface energy balance and evaporation are significantly impacted by these changes in the soil temperature regime. The data on soil temperature from the Minnesota tillage trial for a time after corn or soybean crops were planted are shown in the figure below. 20 0.005-m diam were used to measure the soil's temperature.

Directly buried Type T (copper-constantan) thermocouples at 0.05 m depth in the crop row. By putting four thermocouples 0.76 m apart in five rows towards the middle of each research plot, the thermocouples were set up in a 5-4 grid arrangement. In 2000, when corn was grown after soybean, the three tillage approaches had remarkably consistent average in-row soil temperatures. During the measurement span, daily mean soil temperatures varied by 1 °C on every day and did not exhibit any trend to separate over time or with soil drying.

The soil temperature for the chisel system was consistently 1 °C warmer than the no-till and strip-till systems in 2001 when soybean was planted after corn. With time and soil drying, this temperature differential tended to widen, reaching a maximum difference of 2.45 °C between chisel and no till on Day 190. The 2000 measurement period had more precipitation (199 mm) than the 2001 interval (94 mm), resulting in more wet soil conditions and less soil warming. The majority of the rain in 2001 fell within 30 days of planting, following which soil temperatures rose and the water content of the soil gradually dropped. By tillage system, the daily average, maximum, or lowest soil temperature did not differ significantly in either year. In 2001, when there was no intermediate till, the daily temperature range (max - min) for the chisel system was much greater than for the strip till (mean values for the full measurement interval were 6.39, 5.29, and 5.96 °C, respectively). When maize was planted into soybean residue in 2000, there were likely comparable in-row circumstances that led to the very minor changes in soil temperature between tillage strategies. The soil temperature in the seed zone is strikingly similar for the three tillage systems due to the low amount of residue on the surface even for the no till and strip till systems and similar clearance of residue from the row area during planting. Although not statistically significant, the chisel system resulted in higher soil temperatures in 2001, which were observable and persistent. These results suggest that soil temperature in the intra-row seed zone was not suppressed by the increased amount of maize residue on the soil surface in the interrow area in the no till and strip till systems.

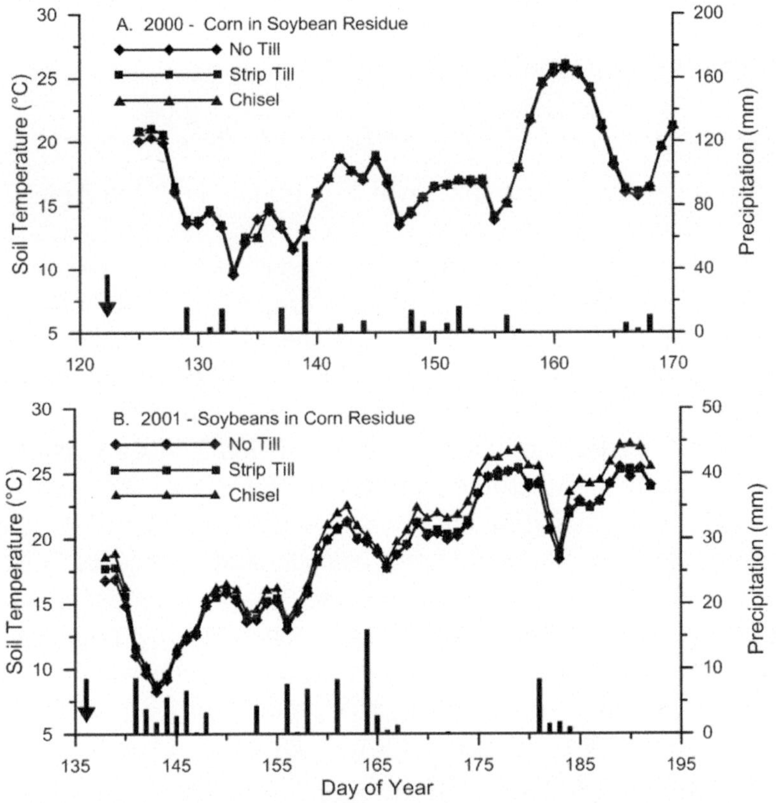

***Figure 31**: Soil temperature data from the tillage trial in Minnesota*

Source: Sauer & Peng, (2020).

In agricultural systems, the characteristics of the land surface and crop canopy influence how incident radiation is divided up and how much energy is required to warm the soil, warm the air above it, or evaporatively cool it. The direction and strength of G can be significantly influenced by the complex interactions between soil and air temperature, soil water content, canopy properties, residue cover, and wind speed. Daytime peak hourly values of G_0 for a bare, dry soil in midsummer, could be in excess of 300 Wm^{-2} (Fuchs and Hadas, 1972; Raoet al., 1977; Enz et al., 1988). Hourly G_0 for a moist soil beneath a plant canopy, residue layer, or snow cover can be less than 20 W m^{-2}. For a wide range of surfaces, such as grasslands, vineyards and orchards, small grains, and row crops with varied canopy cover, soil heat flow data and interpretations are available (Sauer et al., 1998; Hernandez-Ramirez et al., 2009). For non-crop surfaces, such as G beneath sleeping sheep, sparse or desert canopies, or on sloping terrain, further analyses are possible (Oliver, 1992).

Soil Temperature and Heat Flux in Paddy Rice Systems

Wheat (*Triticum* spp.) and maize, which dominate both the American and European agricultural sectors, have been the subject of several surface energy balance studies. There are significantly less surface energy balance studies of rice (*Oryza sativa* L.), another important crop in the world, than there are for arable cropping systems. Over 50% of the world's population is fed by rice, and Asia accounts for 90% of global rice production and consumption (USDA-FAS, 2002). Italy is the top producer, accounting for more than half of the continent's annual production of rice, which is just around 3 million tons. According to IRRI (2009), traditional paddy rice accounts for 54% of the world's rice production and uses around 45% of the freshwater used for irrigation in Asia. Surface energy balance studies for rice agricultural systems are becoming more popular as a result of the demands for expanding rice output (to satisfy the food demand of the growing population) and improving rice water productivity (to meet the growing water shortage in agriculture). The conditions for paddy rice production are more challenging when there are periods of floods, shallow water ponding on the soil surface, and saturated soils. Paddy rice fields occasionally also have the ability to store energy in the paddy water layer (Sw) that is present above the soil's surface. G is consistently found to be the weakest surface energy balance component in paddy rice systems.

According to Harazono et al. (1998), a rice canopy in central Japan under drained circumstances accounted for around 8% of the Rn on average during the day. In the Taichung basin in central Taiwan, Tsai et al. (2007) conducted an energy balance study over a flooded rice field and discovered that the daily average G at 8 cm was less than 1% of the daily Rn. In representative rice-growing regions of Japan, Maruyama and Kuwagata (2010) found that G at the water's surface was roughly 10% of the daily Rn over flooded rice fields. Hossen et al. (2012) recorded G at the soil or water surface (when flooded) in a double cropping paddy field in Bangladesh that was both irrigated and fed by rain. Half-hourly G, with a range of around -50 to 100 W m^{-2} over the day, indicated that G represented just a small part of the energy balance. This diurnal trend was seen during both the summer fallow periods and the flooded cropping periods (with a flooded field and ratoon crop and weeds). In the winter fallow phase, when the field was bare and dry, Rn was equally distributed to G and other components, resulting in a distinct energy partitioning.

The Rn over rice paddy fields could be considerable and include a significant amount of the short-term (i.e., hourly or half-hourly) G. From July 5 to July 10, 2013, rice fields in both continuously flooded and intermittently irrigated rice fields were subjected to half-hour G measurements at the soil or water surface. In the flooded field, the daily peak G values ranged from 200 to 300 W m^{-2} (representing about 35 to 45% of the corresponding peak R_n). The intermittently irrigated field had smaller daily peak G values (from about 170 to 200 W m^{-2}), which accounted for about 25 to 30%of the R_n. Even while G had a lower magnitude at night, it nevertheless contributed the majority of the surface's energy. Harazono et al. (1998) observed that for a rice field with a standing water layer of about 0.05 m, ΔS_w was as large as G_0 and the ratio of daytime average $G_0$to R_n increased to about 12% after adding the ΔS_w. Tsai et al. (2007) found that the average daytime G at 8 cm represented 26% of the average R_n and the average ΔS_w was 5% of R_n. Hossen et al. (2012) reported that the energy balance closure in the flooded period over a double-cropping paddy field was improved by 5% by considering the ΔS_w. Masseroni et al. (2014)observed the ΔS_w in a flooded rice field ranged from about -100 to 150 W m^{-2} in the early rice growing period and there was a phase shift between the daily peak G (measured with heat flux plates 8 cm under the soil surface) and Rn.

Rice growth stage and irrigation pattern have an impact on the microclimate (such as soil and standing water temperatures) and surface energy partitioning over a rice paddy area. In the Philippines, a rice field that was flooded and a rice field that wasn't flooded were contrasted in terms of their seasonal fluctuations in soil temperature (at a depth of 5 cm below the soil surface). The daytime soil temperature in the flooded area was much lower than that of the unflooded field with a standing water layer 3 to 5 cm deep, while the opposite connection was seen at night. In a paddy field that was occasionally flooded, Miyata et al. (2000) found comparable results. This is due to the greater C_s in a flooded field, which makes it warm and cool more slowly than a non-flooded field. The soil temperature during the daylight in the flooded field rose more quickly and was higher than in the unflooded field after it was drained. This is because the soil in the flooded field is darker and contains reduced compounds with low C_s(compared with its oxidized species in non-flooded field (Dean, 1999). The daytime rises in soil temperature were higher and more rapid in the flooded area because it absorbed more solar energy. According to Maruyama and Kuwagata (2010), rice growth decreased the range of daily average G from -25 to 25 W m^{-2} to roughly -10 to 10 W m^{-2}. Similarly, the difference between the standing water layer temperature (T_w) and air temperature (T_a) reduced with rice growth gradually from 2 to 3 °C ($T_w > T_a$ just after transplanting) to a negligible amount before maturity, which was mainly attributed to the decreased solar radiation transmissivity of the rice canopy with increased leaf area index (LAI).

In a study by Jia et al. (2010) on the surface energy balance of a rice field in northeast China, it was shown that the daily mean G at 4.5 cm to Rn ratio decreased as rice growth progressed (from 0.14 before transplanting to -0.08 immediately after harvest). Greater LAI and canopy conductance, which resulted in increased incident energy partitioning into LE rather than G. According to Masseroni et al. (2014), half-hourly G at the soil surface in a non-flooded field ranged from -100 to 200 W m^{-2} in the early rice growth stage (LAI 2), which is much less than G at the water surface in a flooded field (range -150 to 250 W m^{-2}). The variance of G in both rice fields was reduced to a similar range of roughly -100 to 120 W m^{-2} throughout the late rice growth period (LAI > 4) with extended vegetative cover.

A critical component in previous studies for rice agricultural systems is the lack of information on heat storage changes between the soil surface and the G measurement depth (ΔS_s) when calculating the G_0 from G. Failure to include this storage term may result in inaccurate G_0, and thus, poor surface energy balance closure. Ochsner et al. (2007) showed that when the reference depth is sufficiently deep to permit accurate heat flux measurements, heat storage is too large to neglect. As for the flooded rice field with a standing water layer above the soil surface, the high water content in the near-surface soil layer can lead to large soil C_s and thus large ΔS_s. Tsai et al. (2007) showed a daytime average ΔS_s of 40 W m^{-2} (representing 12% of the R_n) over a flooded rice field. However, many surface energy balance studies performed in flooded rice fields ignored ΔS_s when calculating G (Harazono et al., 1998; Maruyama and Kuwagata, 2010). Hossen et al. (2012) did not consider the heat storage between the soil surface and soil heat flux plate placement depth (5 cm), and they suggested this omission contributed to the low surface energy balance closure (about 75%) over a flooded rice field.The interaction of large-scale meteorological factors and surface conditions has an impact on the climate close to the ground. The regional climate has a significant impact on crop growth and production, whereas the local microclimate is influenced by energy exchange within and above the crop canopy. The crucial contact for the transfer of energy between surface and subsurface processes is the soil surface. Because soil qualities affect how incoming energy is distributed between warming the soil, evapotranspiration, and warming the air above the ground surface, soil management methods can have an impact on the climate close to the surface. The rate of all biological, chemical, and physical processes necessary for plant growth and the supply of ecosystem services are significantly influenced by the soil temperature, which is a result of this energy partitioning and the soil thermal regime. The consequences of global climate change on temperature and precipitation patterns have a substantial influence on the conditions of the surface soil, which in turn have important repercussions for crop development and food security. The crucial function of soil in agricultural systems in the context of the water, food, and energy nexus is still a topic of ongoing debate on a global scale. The development of crop and soil management strategies to adapt to a changing climate is made possible by a better understanding of soil temperature and heat flow in various cropping systems. To offer the knowledge and data required to optimize soil management for future farming systems, continued advancements in measuring methods and models of soil heat flow are required.

CHAPTER

5

Water and the Hydrological Cycle in Agriculture

CONTENTS

OVERVIEW

All ecosystems, but especially forests, which often develop in environments where precipitation brings more water than plants can use or soils can store, depend on the hydrologic cycle. The extra water helps streams flow, which meets agricultural needs and urban needs far from the precipitation source. The hydrologic cycle is significantly influenced by vegetation, as the graphic below illustrates. Water influences almost every process that takes place in plants. Water serves a significant function in four distinct processes, though:

- Water is the major constituent of physiologically active plant tissue;
- Water is a reagent in photosynthesis;
- Water is the solvent in which salts, sugars and other solutes move from cell to cell, and from one part of the plant to another;
- Water is an essential element for the maintenance of plant turgidity (hydraulic inflation) necessary for cell enlargement and growth.

Water availability has an optimal level, just like most biological circumstances do. Both too much and too little water might impede the development of the aforementioned activities. An illustration of this is when rapid rain causes cells to burst after a protracted dry spell. Because evaporation is no longer occurring at the original pace due to the roots' hard work in extracting water from the soil, there is an excessive buildup of cell turgor. Two reliable generalizations could be made when examining the physiological relevance of internal water relations on crop yields (Slatyer, 1969):

- The growth and development of most crops can only proceed fully unimpaired, and crop yield is only maximal when high water status is maintained throughout the life of the crop;
- The harmful effects of water deficits are usually most pronounced in tissues and organs which are in the stages of most rapid growth or development.

According to the second hypothesis, crop yield is susceptibe to water stress changes through time in terms of economics. In this perspective, it is quite intriguing to observe that it is common practice on farms to apply stress to various crops at specific developmental phases in order to influence quality development and improve the economic yield—an fascinating tension between quality and quantity. Slatyer's initial generalization is seriously questioned by other over-irrigation risks that exist. Aside from the physiological challenges in determining effects, it appears that correlations between water use and crop yield can only be straightforward in arid regions of the world.

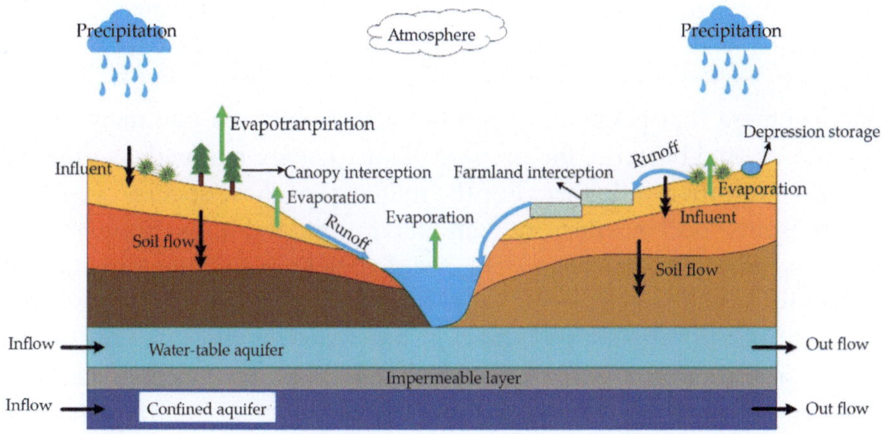

Figure 32: *The hydrological cycle.*

Source: Wang J, Lv X, Qin T, Ni Y, Ma L, Zhang Q, Nie H, Lv Z, Li C, Zhang X, et al. Recognition of the Interaction Mechanisms between Water and Land Resources Based on an Improved Distributed Hydrological Model. Water. 2023; 15(10):1946. https://doi.org/10.3390/w15101946

Despite the wide variety in shape, all forms of vegetation have one thing in common: yield control. This quality is crucial in regard to water. Three components make up vegetative cover: the canopy of living and dead stems and leaves that rises above the soil, the buildup of dead and decaying plant matter that lies on or in the soil surface, and the subterranean stems and roots that permeate the soil. Between the atmosphere and the soil, there are barriers created by the canopy above the ground and the buildup of surface litter. They impede wind movement near the soil surface and deflect precipitation, regulating the amount of water delivered to the soil. Water evaporation from the soil is slowed during the day and may even increase at night due to the heat-insulating qualities of canopy and litter that limit the exchange of energy between soil and atmosphere. The loss of soil water caused by the transpiration of the canopy's living components, however, may cancel out the impact of the canopy and litter. Additionally, vegetation at the soil surface prevents soil material pickup and transport as well as the overland flow of water. This is performed by roots and decomposing organic matter that bind or impede the soil, as well as by stems and trash accumulations that create tiny grounded dams and diversion walls. Vegetation has both direct and indirect effects below the soil surface, down to the depths that roots can reach. By expanding and then dying and decomposing, roots inject the soil with substances that bind it and facilitate the passage of water through it. Additionally, roots take up water and transport it to a plant's above-ground components, which helps the return of water to the atmosphere. By fostering an environment that is conducive to the activity of small animals, insects, fungi, molds, and bacteria, the parts of plants that are below the soil surface indirectly exert their influence. For the majority of these species, vegetation also represents food, shelter, and conducive life conditions. Some of them, such as worms, burrow through the soil and help to maintain its water permeability. Others provide the same kind of role by decomposing organic matter while also infusing the soil with their own bodies' natural binders. Water is absorbed and evaporated from the litter layer and vegetation's surface before precipitation reaches the soil. The density and depth of root canals as well as the amount of organic matter that has been absorbed into the soil determine the pace at which water infiltrates into the soil, flows off the surface, or percolates through to the water table.

Location	Water potential (MPa)	Pressure potential (MPa)	Osmotic potential (MPa)	Gravity potential (MPa)
Outside environment (RH=50%)	−95.2			
Leaf xylem (at 10 m)	−0.8	−0.8	−0.1	0.1
Vacuole of mesophyll (at 10 m)	−0.8	0.2	−1.1	0.1
Root xylem (near surface)	−0.6	−0.5	−0.1	0
Root cell vacuole (near surface)	−0.6	0.5	−1.1	0
Soil (near root)(at −10 m)	−0.5	−0.4	−0.1	−0.1
Soil (10 mm from root) (at −10 m)	−0.3	−0.2	−0.1	−0.1

Figure 33: *Schematic diagram of typical water potential at different locations during water transfer from soil to the atmosphere.*

Source: Ma Y, Liu H, Yu Y, Guo L, Zhao W, Yetemen O. Revisiting Soil Water Potential: Towards a Better Understanding of Soil and Plant Interactions. Water. 2022; 14(22):3721. https://doi.org/10.3390/w14223721 Ma Y, Liu H, Yu Y, Guo L, Zhao W, Yetemen O. Revisiting Soil Water Potential: Towards a Better Understanding of Soil and Plant Interactions. Water. 2022; 14(22):3721. https://doi.org/10.3390/w14223721

On vegetation, precipitation that comes in the form of rain, snow, or fog is intercepted; the amount intercepted varies somewhat depending on the leaf area index (LAI). Throughfall and stem flow enable some precipitation to reach the ground's surface, where it wets the litter, flows off, or seeps into the soil where it may eventually reach the water table. The soil matrix contains some water that is held against the gravitational pull and is accessible to plant roots. Although some of the water absorbed by plants may be temporarily held in the stem, branches, and foliage, the majority flows from the leaves into the atmosphere very quickly. When soil and plant surfaces are moist, water also evaporates from those surfaces. In the example of low stress, the xylem pressure potential gradient from the root to the leaves is about 0.8 MPa, but it is only about 0.2 MPa in the case of high stress, quantifying the reduced water flow rates that occur under high stress. Leaves under high stress can have a pressure potential of 0.0 MPa, causing wilting.

Research implies that vegetation influences water output by affecting water losses through evaporation and erosion processes, slowing or rerouting surface flow, and causing water to enter the soil. Attention must be directed at a thin layer above and below the Earth's surface, especially the layer that sits between the tops of plant crowns and the lowest tips of plant roots, in order to characterize these hydrological impacts in greater depth. The broadest possible interaction between plants and hydrological and climatic processes occurs in this layer. Between being delivered to Earth as precipitation and being released into streams and ground water basins (moving from the realm of meteorology into that of hydrology), water goes through a number of processes, some of which are as follows:

- **Interception:** Precipitation is divided into a part delivered to the soil surface and a part caught on vegetation and evaporated into the atmosphere.
- **Evaporation:** Returns water to the atmosphere from water surfaces and the soil.
- **Transpiration:** Returns water to the atmosphere from the soil through plants.
- **Sublimation:** Returns ice- or snowpack-water directly to the atmosphere.
- **Melt:** Delivers snowpack-water to the soil surface.
- **Infiltration:** Divides water reaching the surface into surface flow and soil water.

- **Surface flow:** Conducts water not absorbed by the soil, and water diverted to the surface after absorption, to streams and the sea ('runoff').
- **Accumulation:** Determines the quantity and areal distribution of a snow pack.
- **Storage and drainage:** Retain water and carries it through and below the soil.
- **Erosion:** Provides sediment load in gullies and in channels, by detachment and transport of material on the soil surface.
- **Deposition:** Separates the sediment load from flowing water.

MOISTURE CHARACTERISTICS OF SOILS

The number and distribution of spaces (referred to as "pore spaces") into which water and air (as well as plant roots and animals) can travel vary depending on the kind of soil and the size, shape, and density of packing of its particles and aggregates. Additionally, there are variations in the surface characteristics of different soil materials, particularly in terms of their capacity to absorb and release moisture. These characteristics have a significant impact on the soil's ability to retain and transfer water. Therefore, the amount of water per soil block, per unit weight of "undisturbed" soil, or per unit weight of oven-dried soil is not a suitable measure of the amount of water that is available in the soil. Even though all soil types have exactly the same total soil moisture content S (expressed as a percentage of dry weight), different soil types will release varied quantities of water to crops. Four convenient categories can be used to organize soil moisture:

- Gravitational water, which occupies the larger pores of the soil and drains away under the influence of gravity. An excess of this water can damage plants if drainage is too slow, for instance by choking the roots;
- Capillary water is held by surface forces as a film around the particles, in angles between them and in capillary pores. In the form of liquid, capillary water moves slowly from thicker to thinner films and along moisture-tension or diffusion-pressure gradients. It can also move in the form of vapor. Capillary water is the only important source of water for most cultivated plants;
- Hygroscopic water, which is held as a very thin film on the particles by surface forces. This water is held so firmly that it

can move only in the form of vapor. The moisture remaining in air-dry soil is usually regarded as hygroscopic water and is not available to plants;

- Water vapor, which occurs in the soil atmosphere and moves along vapor-pressure gradients. It probably is not used directly by plants. The movement of liquid water through soils occurs predominantly under the joint action of gravity (unidirectional) and surface-tension forces (all directions), associated with the degree of curvature of water films around and between particles. Accordingly, these forces depend upon the size and the shape, i.e., the geometry of the pore spaces. Other forces, hydrostatic and osmotic, may also operate.

A block of normally "dry" soil will lose some of any water that is added quickly through any relatively significant fissures and channels. The remaining particles, starting with the bigger pore spaces, will tend to push some of the air out of the spaces between them. In general, as a clearly defined "wetting front" moves downward into the soil, a layer that is growing in thickness and holding all the moisture it can against gravity is considered to be at field capacity. Similar patterns can be seen in the drying of soil, which involves a relatively abrupt "drying front" going downward into the soil as evaporation advances from the surface. As a result, the soil block will typically have alternate wet and dry layers following a period of rainy and dry weather. It is important to distinguish between the favorable condition of "saturated" soil when water fills every pore in the soil, and the state of "field capacity," which is generally desirable for agriculture. The subsequent, very limited movement of water in the soil is primarily in the vapor phase and is primarily caused by distillation, which is connected to temperature gradients in the soil, when the moisture content falls below field capacity.

However, plant roots within the block will draw liquid water from the water films surrounding the soil particles they come into contact with. Therefore, it is evident that a soil block with vegetation can lose more water than a block with the same initial water status but no vegetation. The forces holding the moisture films to the soil particles can no longer be overcome by root suction as water is removed from the soil and the moisture content falls further below field capacity. At this point, plants become water-starved and lose their turgidity because the soil moisture content has reached the wilting point. The amount of water that is available to the plant, which is closely related to surface tension forces in the soil, is more crucial for it than the absolute amount of water present or the volume percentage in a unit volume of soil. This 'available water' is specified by the soil moisture potential Ψ, which is the suction energy (as measured with a tensiometer) that roots must exert to extract water from the soil. The appropriate unit for Ψis J m^{-3}, which usually is translated into a unit of pressure called pF, equal to the logarithm (base10) of the (negative) head of water expressed in cm. For practical purposes, the wilting point is reached for all species when the forces attaching water films to the soil particles reach a value of about 15 atmospheres or 1.52 MPa = 155 m water column ➔ pF = 4.2. This corresponds to capillary action in spaces of less than 1 μm width, while most root hairs are at least a few μm in diameter. So, the usual location of the wilting point at pF = 4.2 makes physical and biological sense.

Another important moisture level is field capacity, which corresponds to pf-values between roughly 2.0 and 2.5 in actual field soil circumstances. This measurement is not exact since gravitational drainage of water does not end suddenly. Therefore, field capacity is typically characterized as being at pF = 2.0. (i.e., suction of 104 Pa).

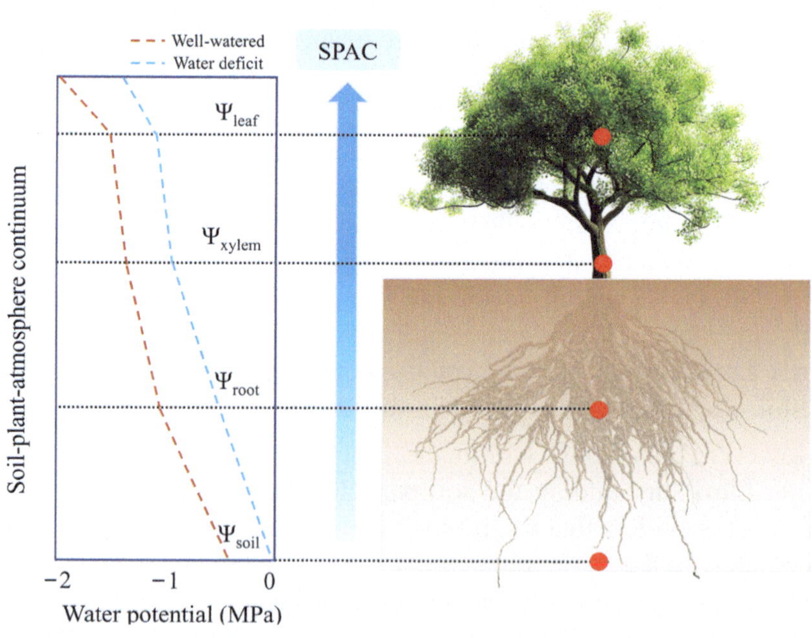

Figure 34: Soil-plant-atmosphere continuum (SPAC) water fluxes*Source: Ma Y, Liu H, Yu Y, Guo L, Zhao W, Yetemen O. Revisiting Soil Water Potential: Towards a Better Understanding of Soil and Plant Interactions.* Water. *2022; 14(22):3721. https://doi.org/10.3390/w14223721*

The volume percentage difference between pF = 2.0 and pF = 4.2 represents the “readily” available water for plant usage, and it strongly changes with soil structure. In this graph, the sand dune, which can hold just 9% of the water that is available, and the peat, which can hold 44% of the water that is available, represent extreme situations. The majority of the water will be withdrawn from the top 0.3 meters of the soil when the water supply is sufficient1, such as during irrigation. More water extraction only takes place from the deeper soil layers when there is a shortage of water (0.6-0.9 m). Rice crops typically have a considerably shorter root system, not reaching 0.5 m, but wheat will often grow a root system that can exceed 1.5 m depth. However, the bulk of the roots are in the near-surface soil layers; the majority of wheat crop roots can be found in the first 0.6 m and the majority of rice crop roots in the first 0.3 m.

DETERMINATION OF WATER LOSS FROM LAND SURFACES

Evaporation is not always measured in mass-balance analyses of a catchment’s water economy; instead, it is taken into account as the last term in a water balance after the other factors (precipitation, run-off, and subsurface flow) have been evaluated. Evidently, it is impossible to make a precise distinction between a number of somewhat ambiguous hydrological processes. The earliest hydrological option for determining real evaporation has been to use the universally accessible meteorological data, such as monthly average temperatures, occasionally with humidity information. Here, we go over several fundamental meteorological options for calculating surface water loss through evaporation from bare soil or open water, as well as transpiration from plants.

FUNDAMENTALS OF THE EVAPORATION PROCESS

Energy, or the latent heat of vaporization, which is approximately 2 500 kJ of heat per kg of water (at T 20°C), is needed for the phase change of liquid water into water vapor. With this energy, evaporation is possible, provided that there is water that can be evaporated and that the resulting vapor is evacuated. There are three stages of evaporation, and each stage is crucial to the process. Capillary rise and distillation are the two basic mechanisms by which water rises in the soil against gravity. When the surface is cooler than the deep soil, soil water vapor rises and condenses once more at the surface.

The efficiency of both processes depends on the soil structure and the gradient in soil temperature. Assuming that water is present at the surface as a result of either of these two processes or recent precipitation, let's examine the following two steps, vaporization and air transport. Three major fallacies about the evaporation process are prevalent in popular beliefs, specifically that:

- High temperatures imply high evaporation;
- Evaporation is proportional to the vapor pressure deficit, or to the relative humidity, as measured in the Stevenson screen;
- Evaporation is proportional to the (daytime average) wind speed.

These statements are at best half-truths, indicating a correlation but not acausality. In a number of non-trivial situations, they are just plainly wrong. Heat, which is not the same as a high temperature, is required for the second step of evaporation, the phase transition of liquid into vapor. Heat is the total energy of all molecules in an object, while temperature measures the average single molecule's kinetic energy. Compared to a cubic meter of equally warm soil, a cubic meter of air has a heat content that is 4,000 times lower. The main energy source for evaporation is net radiation. This means that during the night, very little water evaporates from the ground. All surfaces are gaining and losing long-wave energy at the same rate in dense shade as well, indicating that there is radiative equilibrium, for example, deep inside a dark forest where solar radiation hardly reaches. Therefore, the upward soil heat flow and the heat of the air itself are the only sources of energy for evaporation. The air will significantly cool in the latter scenario, a similar process to the so-called oasis effect. A lake's ability to absorb the necessary heat from the water below implies that evaporation there can occur both during the day and at night.

Incoming radiant energy must be directed to alternate energy sinks if there is no surface moisture to evaporate, particularly by heating the ground until its temperature is high enough for surface radiation emission to balance the short-wave input. Naturally, the air gets heated over such dry surfaces, but due to the air's poor thermal capacity, it is insufficient to act as a sink. Statement (A) is obviously incorrect in this instance because there is no evaporation despite the high surface temperature. The fact that the vegetation may close its stomata at high temperatures, preventing evaporation, is another reason why (A) is unreliable. When the surface is wet and the air is dry, evaporation is stronger than when the converse is true. There are erroneous ideas about how this is measured, though. It is a deceptive half-truth to claim that "the driving force (of evaporation) is the vapor pressure difference between the surface and the overlying air."

The vapor pressure gradient, which is the difference in humidity divided by a distance, is the true indicator of the humidity difference that drives evaporation. It may seem unimportant, but if a given (low) vapor pressure e(a) is measured in a Stevenson screen that is 2 m high (as is typical in France or Norway), the evaporation is likely to be only 2/3 of what it would be if the same e(a) was measured in a screen that is 1.2 m high (as is typical in England), all other factors, such as radiation and wind, being equal. The ecological practice of omitting any mention of the humidity observation height in this situation is blatantly misleading. It is accurate to say that evaporation is inversely proportional to the difference [e(s)-e(a)], where e(a) is the vapor pressure at the surface and e(s) is the evaporation rate at a certain screen height. The difference between the actual and saturation vapor pressure at a single level is how this formula is most frequently understood, which is as a vapor pressure deficit. This inaccuracy may result from misinterpretation (interpreting "s" as "saturation" rather than "surface") or from the false assumption that the saturation point of the surface vapor pressure always exists. For instance, if a somewhat dry breeze blows across extremely dry sand, condensation may happen at the sand surface rather than evaporation. Only the strength and direction of vertical gradients in water vapor pressure can be considered a "driving factor" for evaporation. Following misunderstanding (B) blindly could result in biased or simply incorrect answers. The evacuation of the water vapor by the atmosphere is the third stage of the evaporation process. The movement of the air provides the necessary mechanism to aid in the transport of vapor, but turbulence rather than average wind speed. For instance, a given average wind speed will result in more movement across uneven land than over smooth, flat terrain. Additionally, it indicates that even though the hourly average wind is very light during strongly radiative noontime circumstances, the transport will be substantial due to the tiny convection eddies. Evaporation will be relatively intense in the turbulent wake behind a windbreak of height H, typically at downwind distances of between 2 H and 12 H. Many evaporation equations, especially those of the "bulk aerodynamic" type, as well as the majority of the commonly used combination kinds, feature an aerodynamic term that rises linearly with average wind speed. This can result in erroneously high values of calculated evaporation in areas with consistently high climatological wind speeds (like north-west Ireland). Qualitatively, it seems obvious that further increases in wind speed won't have any impact on rising evaporation above the level of physiologically viable water transport in the crop once there is enough wind speed to generate a violent turbulent exchange activity inside and above the crop stand. When the average wind speed rises above 5 to 7 m s^{-1} for short crops, evapotranspiration is unlikely to increase. The 'maximum effective wind speed' upper limits, however, vary depending on the kind of the evaporating surface since wind speed within a crop is not the same as that above the crop. Evaporation is typically measured as the depth of water lost from a unit horizontal surface (analogous to the measurement of precipitation). Thus, if E mm of water is evaporated, the energy required per m^2 surface is LE, where at ordinary temperatures $L \approx 2.5 \times 10^6$ Joule kg^{-1} ($\approx$590 cal g^{-1}). In strongly evaporative conditions, the radiative energy available for evaporation is about 500 W m^{-2}, which means that in such conditions 1 mm of water will take between one or two hours to evaporate.

Methods to Determine Evaporation

In agricultural meteorology, significant focus is placed on calculating the water loss from crops of various types owing to evaporative processes on surfaces with varying levels of moisture. Given the discussion of the three stages of evaporation above, there are four categories under which to classify agricultural meteorology techniques:

- **Surface water loss measurement:** A device's water loss from a pan of water or a surface that is kept moist by a supply of water from a connected "reservoir" is observed;
- **Measurement of energy used for evaporation ('energy budget method'):** Evaporation is recognized as one of the components in the surface energy budget in physically based relationships;
- **Atmospheric vapor transport measurement ('aerodynamic method'):** Physically-based relationships are used, in which evaporation is regarded as a process of diffusion;
- **Combination method:** Combined application of methods (b) and (c) ('combination method'). The joint application of (b) and(c) eliminates the need for hard-to-get surface information.

Every technique also has a basic operating version. Care must be made to empirically relate these trials to conventional climatological observations in order to establish these results, which will require thorough experiments to determine evaporation at well-equipped research sites. Frequently, only similar climates can benefit from the operational equations obtained. Here, we'll look at method (a), which involves measuring actual water losses from a test body. This method of measuring evaporation appears simple, yet it is deceptive. The key challenge is getting a reading that accurately describes the space or circumstance that the sensor is meant to observe. Lysimetric is a term used to describe a method of measuring water loss directly by carefully observing a section of the soil is actually analyzed. The challenges brought on by the disparities in exposure between the test body and the represented area are only moderately problematic for this procedure. However, even for the most basic form, monitoring the change in weight of a sample of soil, it would take too much time and money to correctly monitor the real water content of a few cubic meters of terrain (there, too, the question of getting a representative sample still remains). In some lysimeter types, provisions are provided to keep the surface completely moist, so that evaporation is not constrained by a lack of water. In other cases, the soil block in question is exposed to the atmosphere unhindered, just like the surrounding terrain is. One gauges a "possible" loss, the other an "actual" loss. Using a man-made vessel that is exposed in the field to assess water loss is an alternate method of determining surface water loss. This category includes atmometers and evaporation pans. We still need a way to determine the corresponding water loss from a soil surface given some measured "open water" evaporation from a pan, typically some pan coefficient or correction factor. Reviews of the conversion factors between evaporation from various sources and pan readings, particularly for the US Class A pan, have been reported in a number of publications. A few square meters of open water are not truly representative of most field circumstances, as evidenced by the extremely huge variances in pan coefficients. Pans and certainly smaller atmometers have their own microclimates that are entirely humid and significantly impacted by the presence and physical characteristics of the water container. They are primarily helpful for comparisons.

Energy Balance Estimation of Evaporation

In the case of a crop, the most general energy balance formulation is:

$$Q^* + A = H + LE + G + g + P$$

Four components are operative in almost all situations (arrows indicate the direction of a flux for which the sign is taken positive), namely:

Q* ↓ Net radiation;

H ↑ Flux of 'sensible' heat (generally turbulent) from ground to air;

LE ↑ Evapotranspiration (unless there is no water to evaporate);

G ↓ Flux of heat into the ground.

Other components which may possibly be present include:

- g If we consider vapor transport in a vertical column of atmosphere, there is heat storage in the crop; but, if we only consider the vertical flux at the surface, there is no such storage;
- P Whenever there is vegetation, energy requirements for photosynthesis must be met, yet on a long-term average, this process only consumes about 1% of input energy, leading many to consider it insignificant.
- A ↓ The advection term, or the horizontal divergence of sensible and latent heat, is the second term from the left. At the most fundamental levels of analysis, this phrase causes a lot of misunderstanding. Assuming it to be insignificant at this stage, it will be covered in greater detail below.

On the surface, the general balance equation reduces to:

$$Q^* = H + LE + G$$

(Q*-G) represents the energy available for the fluxes of heat and water vapor.If the ground is completely dry, then E = 0 and H = Q*-G. If the ground is saturated, or the surface is open water, the maximum possible value of E will be E = (Q*-G)/L.

The temperature profiles may be about the same if soil temperature is then recorded at 24-hour intervals. This indicates that even though heat will have entered and exited the soil block after a cycle of temperature change lasting 24 hours, the overall effect may be nearly nil (G→O). If G is 'made zero' by choice of a 24-hour analysis period, or neglected in situations where it is indeed relatively small (e.g., low soil heat conductivity), then the upper limit of evaporation is(Q*/L).

However, It is not always safe to assume that G is negligible. There are several circumstances when the soil block's temperature profile will differ noticeably at the start and end of a day, a week, or even a month. Such fluctuations could, for instance, result from variations in snow cover over time, the percolation of water from snow melt, or sharp shifts from extremely dry to extremely wet conditions as a result of heavy rain, or the opposite transition from wet to dry or simply from the onset of spring or fall. The soil temperature profile will roughly remain the same throughout the course of a year (measured at the same time of day), and once more G→O. Budyko, quoted in Sellers (1965), used this to achieve maximal simplification. He first found that if Q* is the annual net radiation for a wet surface with an assumed albedo of 0.18, then:

$$Q^* = 10\sum T$$

$\sum T$= sum of daily mean temperatures greater than 10°C.

Next, he argued that, if it is assumed that the convective heat loss H is zero, then:

$$E_0 = {Q^*}/{L} \approx 0.2 \sum T \; mm/yr$$

Observations reveal that when there is no shortage of water, a large fraction of the energy is used for evaporation and a comparatively small amount is used to heat the air. H takes on the primary responsibility of dispersing incoming radiation when water is scarcer. G is a significant factor in the heat balance over water bodies because, in this situation, convective motion inside the water surface layer increases the downward heat flow.

Table 13: *Dominant factors in dissipating energy from different surfaces*

Kind of site	**Site energy budget**
Lush growing crop Meadow	E. is 80 to 90 percent of the dissipation. The largest dissipation mode during growth. When the grass is cut and mats to form an insulator and a vapor barrier (litter effect), then H dominates.
Rocks, City pavement, Desert	E is almost nil; H and G are comparable, depending on the soil heat conductivity.
Open water	E and G completely dominate H.
Still water	G is reduced as compared with open water, since convective mixing is reduced. So, the water surface becomes warmer, increasing both E and H.
Snow (mid-latitude), T < 0°C	Q* is rather small because of large albedo. G is very small (excellent insulation), E is very small (low vapor pressures), so H is a large fraction of Q*.

Source: (Lowry, 1967).

Some of the challenges that are introduced as soon as measurements are taken above the surface are illustrated by a quick analysis of the energy and water budget of an air layer close to the ground. There won't be any change in the temperature and humidity of the atmosphere in the horizontal direction, given a uniform extended area with a homogeneous plant cover, a uniform bare soil surface, or a uniform extended open-water surface; this implies that there won't be any horizontal divergence of sensible or latent heat, and there won't be any advection of energy. Some implementations of the "combination equations" and various energy balance formulae used for estimating soil moisture and irrigation needs do not assume advective effects. The phrase "the area is homogeneous at huge distances surrounding the point of interest" is frequently used to describe this. The vertical fluxes of sensible and latent heat change slowly with height when the advective factors can be ignored, and these vertical fluxes can be approximated from observations taken close to the ground. Deriving LE and H at the surface requires observations, but doing so is actually exceedingly challenging. In order to minimize the impacts of buoyancy and advective effects, the depth of the air layer involved is kept as small as possible. H and LE are typically generated from measurements of temperature and water vapor at two levels (at least) close to the ground. Two issues arise, nevertheless, when the observations are only made at low levels. For starters, even when vertical temperature and humidity gradients are substantial, determining them over a distance of no more than a few meters necessitates the employment of high-quality instruments and expert analysis, which is a difficult goal for station use. Second, at very low elevations above the vegetation, the veracity of several widely accepted interpretations of boundary-layer behavior is open to question. According to Garratt (1978), the minimum elevation profile should be at least three times as high as the roughness obstacles. This need poses little difficulty over low vegetation, but it hinders its application over tall crops, let alone above woods. There must be a second connection between H and LE. In the hypothetical situations:

$$\beta \equiv \frac{H}{LE} = \gamma \frac{T_1 - T_2}{q_1 + q_2} \left(= \gamma \frac{\Delta T}{\Delta q} \right)$$

β is the Bowen ratio, and γ the psychometric constant (here $\gamma = C_p/L$, since we use specific humidity q instead of the vapor pressure e which is used in psychrometry).

Divergence becomes significant where two surfaces meet (the so-called "edge effect"), as in the extreme cases of water to land and arid desert to irrigated field. Conditions can become uniform when they are sufficiently removed from the crop's boundary; typically, this distance is in the range of several hundred meters. Thornthwaite and Mather (1955) proposed a buffer zone with a minimum diameter of 400 meters for a comparable situation involving the placement of a pan evaporimeter in a desert environment. A value of the order of one vertical unit per 30 to 100 units of horizontal fetch is suggested by other research on the development of a boundary layer fitted to the new surface. Slatyer and Mcllroy (1961) point out the possibility of unexpected advective affects even with observations at the surface and/or with horizontal homogeneity. During the day time the net radiant flux (or more strictly Q*-G) is often thought to represent the total 'available' energy, which must be divided up between sensible and latent heat fluxes into the atmosphere. It has therefore been suggested that average Q* must set an upper limit to average LE, even for a thoroughly wet surface. However, the symptoms of H and LE may differ in situations with severe advection. For example, in the case of the irrigated "oasis," strong evaporation occurs at the moist surface, consumed as latent heat noticeably more than the net radiant energy available at the time. During this time, heat may actually be given to the ground from the air for much of the time. Average latent heat consumption will still occasionally be found to exceed average net radiant energy received in any relatively homogenous moist area that is large enough to completely rule out local or low-level advection, frequently over months at a time. This results from what is essentially large-scale or upper-level advection caused by weather system movement, in the case of the evaporating surface, from the passage of relatively warm, dry air masses originating over huge, relatively hot, dry surfaces elsewhere. Due to the general pattern of circulation over them, it appears likely that some regions will act as long-term sinks of sensible heat even though they are quite well supplied with incoming radiant energy - provided, of course, that there is enough moisture available for the amount of evaporation involved. Strong evidence of this impact has been found in Aspendale, Australia, and to a lesser amount in Davis and other locations throughout the United States, as well as in Israel and England.

Aerodynamic Estimation of Evaporation

In contrast to the energy balance approach, measuring the removal of evaporated water by atmospheric motion has the advantage of not requiring data on the condition of the surface at ground level. According to the fundamental tenet of the aerodynamic approach, vapor transfer will happen along a vertical moisture gradient and with an efficiency that is mostly based on the strength of vertical turbulent motion. The most basic way to use this process is to measure the actual turbulent air movements that cause eddy diffusion while also monitoring the related changes in humidity that determine how well the air motions transport water vapor. The experimental use of this basic methodology is known as eddy flux or eddy correlation observation. It necessitates measuring at high frequencies of 1 Hz or faster over both the vertical components of the wind and humidity changes. This can be done via hot-wire, sonic, or fast-propeller anemometry, but the work and expenses are so high that operational use is not really practical (as with lysimetric, another "basic" method). However, by focusing on the fundamental tenet that the general amount of turbulence is an useful indicator of the effectiveness of air transport, the method can be made simpler and more suitable for operational reasons. The measurement of vertical wind components can therefore be discarded, and the evaluation of a complete "high-frequency" time series is substituted by the more straightforward calculation of wind or temperature standard deviations. This has been useful for fluxes of sensible heat as well as boundary-layer wind analyses. We are now working on approaches that may be used in the field to calculate evaporation from turbulent flow at the mesoscale.

The aerodynamic profile approach is based on the fact, that the turbulent exchange is linked to the vertical wind speed gradient, i.e., to dU/dz. Thornthwaite and Holzman (1939) were the ones who initially devised this, and even though their equation is the least accurate (it is only "valid" under neutral stratification), it is still the only one used in textbooks because it was the first and is the simplest to calculate. More dependable strategies have replaced the formula. Deacon and Swinbank's additional, theoretically equivalent but more flexible wind-humidity profile formula for evaporation;

$$E = -\rho C_{D(z)} U_z^2 \frac{q_1 - q_2}{U_1 - U_2}$$

where ρ is air density and $C_{D(z)}$ is the drag coefficient, a surface roughness measure related to the roughness length z_0 by $C_{D(z)} = (0.4/\ln (z/z_0))^2$, where $C_{D(z)}$ and $U_{(z)}$ should refer to the same height. The dependence of $C_{D(z)}$ on thermal stability is known, so Deacon-Swinbank is much more applicable thanThornthwaite-Holzman.

However, due to the gradual discovery of the magnitude of the uncertainty in short-mast profiles of wind, temperature, and humidity in non-homogeneous terrain, aerodynamic profile methods like these are now less frequently used in the field. Unless upwind topography is uniform over distances of around 1 km, profiles cannot accurately reflect turbulent fluxes for evaporation estimation (Wieringa, 1993). It is preferable to quantify gustiness or variation rather than use profiles if you need accurate flow information. Formally speaking, the bulk aerodynamic approach is a simplification of the aerodynamic profile approach. For example, when we take the lowest profile level at the surface ($z_1 = 0$) in the Deacon-Swinbank formula, we retain a formula of the type:

$$E \approx constant \times (U - function) \times (q_{air} - q_{surface})$$

Evaporation is proportional to the vertical water vapor gradient and to the wind speed, involving, of course, a proportionality constant. This was Dalton's (1802) first consistent explanation of the evaporation process. There are numerous formulae of this type that represent wind and vapor pressure in all conceivable types of units. All of them are based on the fundamental notion that the movement of water vapor from a surface is caused by a mixing process (particularly, turbulence caused by the wind) that is active between two levels with varying vapor pressures (or vapor densities). All of them are based on thorough tests conducted in specific terrain and climate settings, using observation heights that were convenient or available nearby. Because the proportionality constant(s) must take into account the relationship between the average wind speed at a locally selected height and the overall degree of turbulent transport, bulk equations are never generally applicable.

A bulk-aerodynamic formula developed for estimating lake evaporation, for example, from observations at 2m height is only directly applicable to lake evaporation estimates in a similar climate from similar data, though it may do that exceptionally well. This is because such a relation is terrain- and stability-dependent. In a strict sense, bulk aerodynamic formulas are not truly aerodynamic. The wind-profile and eddy-correlation approaches to evaporation estimation solely use air measurements, whereas a bulk formula needs data on surface-level vapor pressure, which is difficult to obtain. Because water surface temperatures are simple to measure, bulk equations have thus had the best success in calculating open-water evaporation under the assumption of full saturation at the water's surface.

Combination Methods

The simplified bulk-aerodynamic approach and the energy balance both require knowledge of surface properties like temperature and vapor pressure or specific humidity, which are challenging to measure at the surface with any level of accuracy and do not typically form part of climatological observation schedules. Additionally, profile measurements are needed for the energy balance approach and the true (non-bulk) aerodynamic method, which are again excluded from routine station observation methods. Penman (1948; 1956) took the creative step of fusing the energy balance with the aerodynamic approach to establish a strategy to get rid of the "unknown and unknowable" qualities at the surface level. His original method produced what he dubbed "open water evaporation," which he then connected to the evaporation of wet grass, and only required measurements that are actually made at normal weather stations: temperature, humidity, wind speed, and sunshine duration. This seemed too wonderful to be true, but it required more study to apply his findings to the evaporation of genuine open water or plants. The number of modified "Penman" equations—some created by Penman himself, others by others—has grown to the point that it is currently impossible to determine what is meant by "Penman-evaporation" without knowing the specific formulae that are being utilized. Additionally, a number of simplifications were created, most notably the Priestley-Taylor evaporation formula. Furthermore, Monteith and Rijtema (1965) created a method using transfer resistances that required fewer empirical output translations. This method enables the systematic handling of the more complicated vegetation-related circumstances. The FAO recognized the so-called Penman-Monteith equation as the most accurate method for estimating evaporation from meteorological data in 1990; however, this does not imply that it is always the best practical method to do so.

Penman Equation

In order to represent water loss from the Rothamsted sunken evaporation pan, Penman created his original formula and fitted some of his empirical constants to the pan data. He could safely ignore the soil heat flow G, which is minimal for such a pan as a result of this starting point. Therefore, if we specify below the basic heat input as (Q*-G), this differs from the original presentation by Penman (1948) who only considered Q*, the net radiation at the surface. Penman used the following equations to determine E:

- The energy balance;

$$Q^* - G = H + LE$$

- A bulk-aerodynamic heat flux formula;

$$H = C_p f_H(U)(T_0 - T_a)$$

- A bulk-aerodynamic evaporation formula;

$$LE = f_E(U)(e_0 - e_a)$$

- Assumed vapor saturation at surface level;

$$e_o = e_s(T_o)$$

where T_0 and e_0 are at surface level and T_a and e_a are at screen height (in England: 1.2 m). He made the assumption, generally accepted in micrometeorology, that the turbulent transport functions for the scalars T and e are similar:f_H (U) = f_E (U). An essential point in the formula derivation is now the elimination of the variables at surface level by use of the temperature derivative of saturated vapor pressure:

$$\frac{e_o - e_a}{T_o - T_a} \approx \frac{\partial e_s}{\partial T} \equiv s \equiv \Delta$$

The value of the derivative for a given T_a is a known physical function. The result of Penman's analysis is:

$$LE = \frac{\Delta(Q^* - G)}{\Delta + \gamma} + \frac{\gamma L}{\Delta + \gamma} E_a$$

Here γ is the psychrometric constant and E_a accounts for the evaporative power of turbulence: if (Q*-G) = 0, the bulk-aerodynamic evaporation remains. Generally, E_a is called 'wind function' and written in the form:

$$E_a = f(U)(e_s(T_a) - e_a)$$

It is common to refer to the two terms in Penman's equation as the "energy balance" and "aerodynamic" terms, respectively. It is necessary to get the parameter values from station data, particularly radiation:

$$Q^* = K \downarrow (1 - \alpha) + L \downarrow - L \uparrow$$

If global radiation is not measured, we can infer it from cloud cover or sunshine duration. For Southern England, Penman utilized an Angstrom-type regression of radiation on the sunlight ratio s/S; this took the following form:

$$K \downarrow = [K](1 - \alpha)(0.18 + 0.55\, s/S)$$

Penman used an albedo α= 0.05 for open water, appropriate for his evaporation pan-surface. Subsequent investigators have used a vegetation albedo α =0.25 instead. Incoming long-wave radiation can be estimated according to Brunt as Penman did or (better still) to Swinbank or Idso and Jackson. Emitted long-wave radiation (σT^4) was estimated by Penman directly from Ta; correction is desirable when strong vertical temperature gradients occur. Temperature and humidity are measured in the screen. There is no generally accepted way to obtain the wind function E_a; Penman had wind measurements at 2 m height (acceptable at the open Rothamsted site) and used f(U) = 0.26 (1 + 0.54 U_2). Because "Penman did it," this historical fact has unfortunately been replicated elsewhere.

Wind monitoring at a higher level is required in difficult terrain. Alternatives for the wind function are provided by Brutsaert (1982), along with stability effects. The longer the averaging period required for accurate findings with Penman's formula, the greater the dependency on assumed component values and empirical relationships. It has been proven to be sufficient for daily values when using reliable station data (assuming the heat storage term G is not ignored), but it should not be utilized for periods shorter than around five days if the data quality is not guaranteed. In his groundbreaking study, Penman (1948) contrasted measured water loss from planted lysimeters with his deduced and measured evaporation from his pan. The transition from water to vegetation surface required a correction coefficient of 0.6 for winter, and 0.8 for summer use, to convert his 'open water evaporation' (E_0)to green crop evapotranspiration. He justified his coefficients (Penman and Schofield, 1951) by noting firstly, that transpiration takes place through open stomata; this is a response to light, and hence one which is affected by day length. It was discovered that there is resistance to the diffusion of vapor through stomata. Second, crops reflect more solar energy than open water. The greater surface roughness of vegetation's leaves may have a role in boosting evaporation, which is another important factor to take into account when comparing evaporation from an open water surface to that from vegetation.

These similarities led to the development of the term "potential evapotranspiration" (PET), which was first introduced by Penman and was initially described as "the amount of water transpired in unit time by a short green crop entirely shading the ground, of uniform height, and never short of water." This (widely used) notion is extremely confusing, as correctly noted by Brutsaert (1982) because it includes biological elements like the type of crop and the stage in its growth cycle. If potential evaporation is defined as the evaporation of a broad uniform surface that is sufficiently wet to permeate the surface air, then there may be rather significant disparities between PET and potential evaporation. Furthermore, the projected energy distribution between H and LE may be extremely biased if potential evapotranspiration is derived from measurements made under non-potential conditions.

So, the practical solution, therefore, is: first, we define exactly the type of surface for which evaporation is calculated from meteorological data, and call this grass reference evapotranspiration ET_0 (Allen et al., 1994). Then, we investigate how actual evapotranspiration of various crops in various stages of phenological development relates to ET_0, and call this ratio the crop coefficient; similar coefficients can be determined for unplanted surfaces. The practitioner has a number of choices for the method to determine ET_0. FAO now supports the big-leaf formula of Penman-Monteith, but since not all measurements needed for this method are readily available worldwide, there are alternatives. The FAO gives as alternatives:

- techniques that just consider temperature (Blaney-Criddle appears unreliable; Hargreaves' approach using maxima and minima may be preferable);
- a useful radiation-only method based on Makkink;
- the good old pan evaporation, where that is the only available information source.

Pan-crop coefficient tables are available in the FAO review reports (Doorenbos and Pruitt 1975; Doorenbos and Kassam 1979). The reference hypothetical 'grass' crop of the ET_0 -definition is: height 0.12 m, roughness length 0.015 m, displacement height 0.08 m, albedo 0.23, and surface resistance r_s = 70 s m-1. Like the crop of the PET-definition, it is assumed to be 'actively growing, completely shading the ground and never short of water'. If an actual crop is tall and moist, its actual crop evapotranspiration (AET) can exceed ET_0; however, it is usually $AET <= ET_0$, either by reason of the plant's control over the stomata openings, or the failure of the physiological mechanism of moisture transmission in the plant, or exhaustion of the soil water available to the plant. The latter circumstance makes irrigation as a supplement to water supply desirable. Methods for measuring the amount of soil moisture that declines to a point at which root system extraction becomes inhibited (i.e., in the measurement of the Soil Moisture Deficit (SMD)) are the subject of various theoretical discussions and practical challenges.

Evaporation formulae of Priestley-Taylor and Penman-Monteith

Vegetation evaporates in stages, starting at the root and moving through the leaf, the stomata, the canopy, and finally the free atmosphere. Our understanding of the transpiration process has been improved with the use of concepts related to the resistance of plants and canopies to the transfer of characteristics, such as vapor flow. According to Ohm's law, the quantity of current in the transmission of electricity across a wire is determined by the voltage (potential difference between the ends) divided by the wire's resistance. We can assume: as an analog for the flow of any other quantity

$$flux\ rate = (concentration\ gradient)/(resistance\ to\ flux)$$

The simplicity of combination is a benefit of this strategy. Resistances in a chain can be added up, and resistance networks can show situations with multiple possible outcomes. Through utilizing concepts of resistance to water flow within the leaf, then at the leaf-air interface, and finally through the crop stand, variations in the mechanism of moisture loss by leaf surfaces have been made possible. In particular, the 'wind function' E_a in the Penman evaporation determination can be made more accessible to analysis by rewriting it as air resistance r_a, which decreases when the wind speed increases: $r_a = (\rho C_p)/(\gamma f(U))$. With this notation, including some physical constants for convenience, the Penman equation reads:

$$LE = \frac{\Delta(Q^* - G)}{\Delta + \gamma} + \frac{\rho C_p}{\Delta + \gamma} \times \frac{e_s(T_a) - e_a}{r_a}$$

where now the 'wind function' is made visible as 'air resistance' r_a. It has been shown that r_a can be derived as follows from the length-dimension parameter for surface roughness, z_0, and a similar terrain parameter for heat transfer, $z_{0v} = 0.2\ z_0$:

$$r_a = \frac{1}{k^2 U_z}\left[ln\left(\frac{z-d}{z_o}\right) ln\left(\frac{z-d}{z_{ov}}\right)\right]$$

Here K ≈ 0.40 is the Karman constant, and d is a correction height interval; overcrops of height H, approximately d ≈0.7 H. Air resistance r_a is small, when the wind is strong and the terrain is rough, so that the turbulence is vigorous.

It is discovered that the value of the ratio between the two components of Penman's equation is comparable for many different types of short vegetation if they are not truly wet but have access to enough water. This might be explained by some compensating interactions between stomata resistance and roughness that causes turbulence. As a result, Priestley and Taylor (1972) reduced the Penman equation to:

$$LE = \propto \frac{\Delta(Q^* - G)}{\Delta + \gamma}$$

The constant α here has in many cases values between 1.25 and 1.30, though smaller values (e.g., for coniferous forest) or larger values (e.g., for winter situations)can occur (De Bruin and Keijman, 1979). Wintertime evaporation from forests is significant due to the relative importance of the aerodynamic component, whereas short-crop evaporation is minimal over the same period. There are also even more extreme "shortcuts" as a result of the Priestley-Taylor simplification. By effectively utilizing the global radiation measurements that are available in the national station network, the Makkink relation, for instance, has been demonstrated to function acceptably in the Netherlands in summer (Makkink, 1957; Hooghart, 1987):

$$LE = 0.65\, K \downarrow (\Delta/(\Delta + \gamma))\, mm/day$$

This equation is used locally with the proper crop factors (Hiemstra, et al., 1994). Makkink operates admirably in dry areas (De Bruin and Lablans, 1998). If radiation is measured but no humidity measurements are available, another variation is the ability to combine the energy balance and profile approaches to predict evaporation. If we can then determine H from temperature profiles and good wind observations, the determination of LE is also possible (Stricker and Brutsaert, 1978). This approach has an interesting future in remote sensing applications, if these can provide good estimates of Q* or K↓ and of H.A fundamental improvement to the Penman approach has been the addition of a bulk stomatal resistance r_s, to model transfer between leaf stomata and atmosphere. This is often called the 'big-leaf approach to account for the active influence of vegetation on its transpiration. Monteith (1965) and Rijtema (1965) showed that this extension of the Penman equation results in:

$$LE = \frac{\Delta(Q^* - G) + \frac{\rho C_p}{r_a}(e_s(T_a) - e_a)}{\Delta + \gamma\left(1 + \frac{r_s}{r_a}\right)}$$

If in this Penman-Monteith equation, we substitute $r_s = 0$, i.e., we assume that the leaves are covered with water, the original Penman equation is recovered. If we take $r_s = \infty$, meaning that stomata are closed and leaves are dry, we get the trivial extreme, LE= 0.

The 'big-leaf' canopy resistance is an integral property of a type of vegetation and must, therefore, be determined experimentally. Some typical values are $r_s \approx 30$ s m^{-1} for low crops and $r_s \approx 150$ s m^{-1} for pine forests. FAO assumes $r_s \approx 70$ s m^{-1} in their definition of grass reference evapotranspiration ET_0, and it is cogently argued that evapotranspiration estimates from weather data using this approach are superior to lysimeter data as benchmark values (Allen, et al., 1994).

The Penman-Monteith equation, which offers a mechanism to take into account different canopy effects, is a notable example of the power of resistance formulations. The primary benefit that it has with the original Penman equation is that it explicitly incorporates two of the three evaporation stages. For the numerous areas lacking simple meteorological information, however, temperature-only evaporation estimations and the evaporation pan will continue to be used. The discussion of evaporation techniques above is a good example of the challenges faced by applied meteorologists. The use of procedures based on basic physical principles is frequently only achievable with precise tools and information that are lacking in actual field operations. This makes it vital to apply multiple methods and as many cross-checks as the investigator is capable of while solving practical field situations. The estimation of evaporation water losses of the Aswan dam lake on the Nile, where data are quite poor, is an intriguing example. Several of the aforementioned techniques are used, and only if all results roughly concur can we state that they likely have some validity (Sadek, et al., 1997).

SPECIAL FORMS OF' PRECIPITATION

Dew

Dew is not often regarded as a type of precipitation because of its short travel distances, which are estimated in meters only as opposed to the kilometers that a raindrop may travel. However, in particular circumstances, dew can significantly contribute to the moisture balance by depositing water where it is desired or necessary. Dew has a variety of advantageous impacts in arid regions. Some plants can absorb water from their leaves as long as it is evaporating off leaf surfaces, the stomata are open, and assimilation can occur. Dew is a significant drawback in that extended leaf wetness might result in fungi illnesses. The prevention or delay of frost is a benefit. Dew comes in three different forms. Dewfall, or atmospheric condensation, is the first. It needs the exact opposite of high evaporation circumstances, including an extremely negative radiation balance (night with a clear sky), no wind, and moist air. In the absence of wind during significant nocturnal cooling, downward vapor movement is insufficient for dewfall; in the presence of strong wind, a thick layer of fog may instead form. The highest radiational cooling occurs in broad leaves, which can already exhibit nighttime condensation at relative humidity levels of about 90% while soil heat flow still delays dewfall at the surface. Dew rise is the second type of dew, which is distilled from the soil. When the vegetation cools off at night after a sunny day, the vapor from the soil surface condenses on the vegetation. In certain circumstances, condensation can start several hours before dusk and continue beyond it. Dew typically starts after dusk and ends around sunrise, though this isn't always the case (Monteith, 1957). Dew that has already developed may still be there in the morning, although it normally dries out once the condensation stops within a few hours. The third type of dew is water movement from another location within the plants, including the possibility of water exuding from plant tissue (a process known as "guttation") as a result of intense pressure produced by the roots. The amount of dew is often modest. Most often, distillation results in deposits that are much smaller than 1 mm. Dewfall occurs more irregularly and can produce up to 1 mm (Jacobs and Nieveen, 1995). At weather stations, these amounts are not usually accurately recorded. In humid nighttime advection zones, dewfall can be very substantial; for instance, in coastal Nigeria, nocturnal advection is responsible for 500 mm of yearly dewfall.

Snow

In the hydrological cycle, snow and ice serve two functions: first, as a source of moisture; and second, by delaying the release of moisture to the ground, rivers, etc. until after precipitation. There are some effects on water supply, budgeting, and control from this delay. Even if we limit ourselves to hydrological factors, there are still some consequences of snow, ice, and frozen ground on agriculture that we can list:

- Snow protects buried crops from dangerously low temperatures and insulates the earth, but it is dangerous to vegetation that sticks up above the snow.
- The structure of the soil may benefit from light freezing.
- When snow is more than 0.1 meters thick, it partially or fully obscures herbage from grazing animals and poses a physical risk to cattle.
- In terms of mobility, frozen ground after a rainy fall provides a stable platform for field and track vehicles, and snow facilitates sledge travel (e.g., of fertilizers and farm produce). The saturation of soil by melt water in the spring may result in a period when it is either inappropriate or impossible to utilize mechanical farm equipment.

- Following snow and ice loading, snow and ice cause stress on all kinds of structures.

Whether these climatic factors are a consistent annual occurrence or only occasionally occur has an impact on how favorably and unfavorably they balance out. Its water production is vital for the hydrological input of snow. It is generally accepted that 10 cm of fresh snow equals 1 centimeter of water. The density of new snow cover is influenced by the air temperature during snowfall.

Table 14: *Air temperature effect on the density of fresh snow cover*

Air temperature (°C)	**Snow density (g cm^{-3})**	
	Mean value	*Minimum value*
-10°	0.07	0.01
10°to 5°	0.09	0.02
-5° to 0°	0.11	0.04
0° to 2°	0.18	0.07
2°	0.20	0.16

Source: Ventskevich (1961)

The density of the cover, which is initially only 0.1 g cm^{-3}, rises with settling, the mass of any new snow deposit, the length of the snow cover, and the alternating partial thawing and freezing. The density may even double in a day. Snow density is 0.3 to 0.5 g cm^{-3} or greater at the start of the spring melt season. Permanent snow cover on mountains becomes hard ice, increasing the density to 0.8 g cm-3. In cold climates, snowfall can last for many months. Since advection of dry air causes the direct evaporation (also known as "sublimation") of water, which requires latent heat and cools the snow surface, the snow disappears primarily by melting under the influence of solar radiation, by sufficient rain (very light falls of rain may merely add to the snow cover), and most effectively by advected moist warm air. The length of the time, which normally lasts two weeks in Russia, depends on the depth and size of the cover. There will likely be sometime between the conclusion of the thaw and the day the soil is suitable for agriculture. Whether the snow cover formed on an unfrozen surface or whether the ground was extensively frozen before the snow cover built up determines the subsequent disposal and transport of melt water. In the first scenario, melt water can seep into the soil; in the second scenario, percolation won't occur until the soil has thawed.

SOIL MOISTURE BUDGETS

The division of soil water inputs into different outputs and storage is quantified by its water budget. The nature of the physical, chemical, and biological activities that take place within the pedosphere will be greatly influenced by the balance between the outputs. For instance, variations in the rate of infiltration and consequent soil water status in sloping landscapes play a major role in the variety of soil properties across catenary sequences. In order to characterize the pollutant flux to groundwater and other phenomena like water available for plant growth, slope stability, and irrigation scheduling, an accurate assessment of a soil's water budget is a crucial first step. According to the figure below, a soil water budget consists of input, output, and storage components.

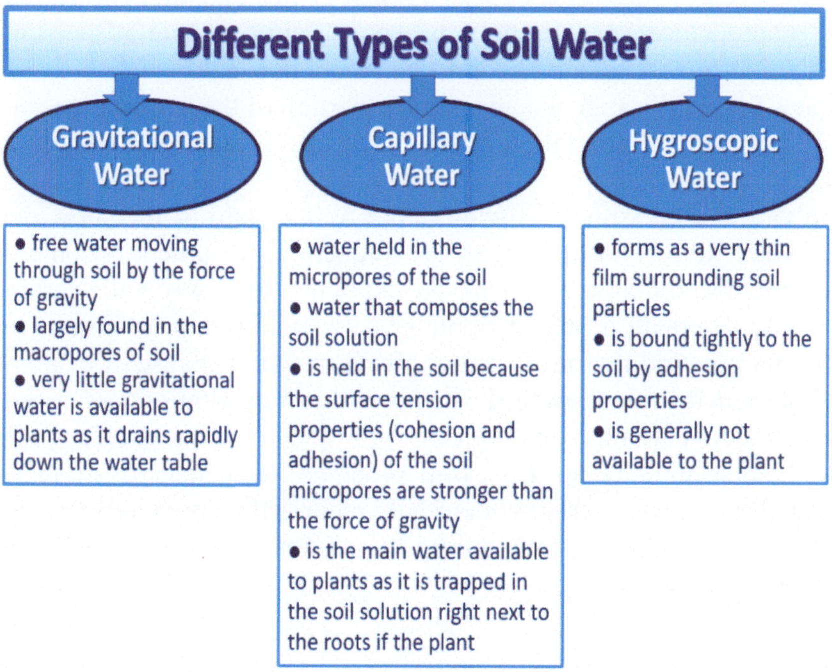

Figure 35: *Relationships between the components of the soil water.*

Source: Gavrilescu M. Water, Soil, and Plants Interactions in a Threatened Environment. Water. 2021; 13(19):2746. https://doi.org/10.3390/w13192746

Precipitation

The two natural inputs for the soil water budget are precipitation, which includes both rain and snow. For many parts of the world, precise precipitation records are available. For instance, the Meteorological Service of Canada, a branch of the Federal Government, provides records in Canada. The National Climatic Data Center in the US gathers and stores precipitation information. In several nations, automated recording stations have taken the place of manual precipitation gauges. A popular technique for automatically and constantly measuring rainfall is the tipping-bucket rain gauge. Raindrops are directed into one of two small buckets that are balanced on a cantilever by a circular-topped funnel with a narrow bottom. One bucket falls and tips, dumping its contents as it fills with water. The second bucket then tips after being filled with rainfall, ending the cycle. Each half-cycle (i.e., tip of a bucket) completes an electric circuit, and each tip is counted by a datalogger while the cycles continue as long as it rains. Typically, gauges are calibrated at the manufacturer to accurately measure 0.2 mm of rainfall per tip. But each tipping-bucket gauge needs to be frequently calibrated to guarantee accuracy. Snowfall can make up a sizable portion of the annual precipitation in frigid climates. Snowfall depth is frequently measured manually using a ruler in the past and even now. The precipitation is then estimated by multiplying the measured snowfall depth by the ratio of the snow density to the water density. A ratio of 0.1 is frequently presumed. A sonic depth recorder can automatically check the amount of snow on the ground. It is possible to create a heated tipping bucket rain gauge to automatically record precipitation that turns into snow. However, because of evaporative losses of melted snow from the heated buckets, precipitation may be greatly understated. to learn more about quantifying snowfall. Snowfall and rainfall are both measured by an all-season precipitation gauge. A strain gauge connected to a datalogger gravimetrically records the precipitation as it falls into a bucket. Snow that falls during the winter can be melted by adding antifreeze to the bucket.

Irrigation

Particularly in dry countries with heavy agriculture, irrigation water may be a considerable financial input. A little amount of water is typically added by other anthropogenic inputs such as liquid manure and sewage biosolids, which can be budgeted for under irrigation. It is quite easy to measure the amount of water sprayed to the soil surface during sprinkler irrigation using an open-topped container. Divide the volume of water captured in the container by the cross-sectional area of the open top to determine the amount of water applied to the soil surface. A technique for determining soil water status is necessary to determine how much water actually percolates into the soil following an episode of precipitation or under dripper or sprinkler irrigation systems. A popular technique for measuring the volumetric water content of soil in the field or a lab is time domain reflectometry (TDR). Topp et al. introduced the TDR method to the field of soil science to measure the volumetric water content of soil (1980). It offers many benefits over other field techniques, including the following:

- does not require a radioactive source;
- high level of accuracy over a wide range of soil types using a single calibration curve such as Topp et al. (1980) equation;
- consistent sample volume size regardless of soil water status;
- nondestructive, so measurements can be repeated at a site; and

can be set up for remote, and automated data collection.

The increase in local soil water storage (mm) during irrigation is simply the measured volumetric water content multiplied by the TDR probe length when employing TDR probes inserted vertically in the soil beginning from the soil surface. When steady-state water content is detected by the TDR system, for instance, irrigation can be stopped if the length of the vertical TDR probes is equivalent to the depth of the plant root zone. Commercially accessible automated irrigation control systems based on TDR and other electromagnetic devices improve water use efficiency. See Topp and Ferre for a more detailed analysis of this and other techniques for determining the water content of soil (2002).

Evapotranspiration

A soil water budget's primary output component is frequently evapotranspiration. It includes water that is used to grow plant tissue as well as water that evaporates from bare soil and wet plant leaves and returns to the atmosphere (Tindall and Kunkel, 1999). The rate of evapotranspiration is influenced by both climatological elements (air temperature, wind speed, and solar radiation received) and the water content of the soil. To calculate potential evapotranspiration using climatological data, numerous numerical techniques are available (Jensen et al., 1990). To accurately predict actual evapotranspiration, these approaches must take varied crop coefficients and the amount of water in the soil that is available into account. The price of meteorological instrumentation has recently decreased, making the eddy-correlation method a practical way to measure evapotranspiration (Verma, 1990). See Brutsaert for further details on this precise, direct method of calculating evapotranspiration above a variety of plant canopies (1982). A regional water management strategy plan must include an accurate estimate of the evapotranspiration requirements of horticultural and agricultural plants (Rudra et al., 2000). Accurate evapotranspiration measurements are essential because of the growing competition for freshwater and the need for good scientific data to support water management policies.

Drainage

In contrast to negative drainage, which reflects a net gain of water, positive drainage indicates a net loss of water from the soil profile. Positive drainage is the movement of water through soil macropores and/or smaller pores past a depth that corresponds to the local soil profile thickness. It can occur during irrigation and precipitation events. The maximum depth of pedogenesis, a random depth, like 1 m below the surface of the ground, the depth of the local plant root zone, or, in humid climate zones, the depth below the water table, are all examples of local soil profile thickness. Positive drainage may also happen over a few days after rain or irrigation stops due to the downward redistribution of soil water. As pressure gradients diminish and soil water becomes relatively evenly distributed with depth, especially in uniform soil profiles, gravity eventually replaces all other forces dragging soil water downward. Water that enters a soil profile from below its lower limit as a result of plant root or evaporative water demands surpassing the soil profile's available water is known as negative drainage. Capillary pressures, which are caused by higher to lower soil water pressure heads from the bottom to top of the soil profile, draw water upward against the force of gravity.

Drainage is one of the most difficult parts of the soil water budget to quantify. It can be done so either directly or indirectly. Lysimeters and a freshly created soil water flux meter are tools for measuring local drainage directly. Using Darcy's Law to determine the downward soil water flux is one indirect way to measure drainage. Two variables are needed to calculate Darcy's Law: the local unsaturated hydraulic conductivity and the hydraulic head gradients in the soil profile, which are measured with tensiometers buried at least two depths down. As a result of the geographical and temporal variability in pore properties, it is challenging to measure the unsaturated hydraulic conductivity of soil, even with an order of magnitude accuracy. Lysimeters can be split into two types: weighing and non-weighing, albeit having a wide range of sizes and designs. Lysimeter weighting is more expensive and typically involves repacking and uprooting the soil profile. Non-weighing lysimeters can be made to collect drainage water under zero or non-zero tension lower boundary circumstances, and their installation is significantly simpler and less disruptive. Before water may enter the drainage collecting device located at the base of a zero-tension lysimeter, a saturated zone must first form above it. In arid places with significant unsaturated zones, this perched water table may result in extremely abnormal soil water flow conditions.

With wet soil and shallow groundwater, zero-tension lysimeters are therefore better suited for collecting drainage. It was successful for Brye et al. (1999) to create an equilibrium tension lysimeter. In comparison to similarly sized zero-tension lysimeters put in other soil systems, they found lesser variability in measured drainage between their sampling locations. To apply the tension, their technique does, however, need a vacuum pump. Gee et al. (2002) made a soil water flux meter that appears to be a very promising approach for measuring drainage effectively and precisely. As opposed to Brye et al., their flux meter concept applies tension to the soil using a passive capillary wick system and does not require a vacuum pump (1999). In order to measure the amount of pollutants that are transported to groundwater, drainage is a crucial part of a soil water budget. For this reason, it is crucial to make an accurate estimate of the quantity and timing of drainage in order to inform decisions about agricultural nutrient management practices, for example.

Overland runoff

Overland runoff is the extra rain, snowmelt, or irrigation water that falls on land but does not percolate into the local soil profile. Instead, it either runs into surface waters or transforms into run-on-contributing water for another soil profile at a lower elevation. A number of variables, such as the percent slope and surface roughness of the landscape, the type of vegetative cover, the soil's sorptivity, the intensity and duration of the precipitation event, and many others, affect how much water runs off a land surface during a specific rainfall, irrigation, or snowmelt event. It is technically necessary to direct runoff water into a well or sump situated at the lowest point in a field in order to measure overland runoff. The amount of overland runoff (mm) is equal to the volume of water collected in the well or sump divided by the field's area. An important factor in calculating soil erosion losses from a landscape is overland runoff.

Subsurface Runoff

Water that flows down a sloped terrain inside the depth of a soil profile—also known as runoff—is referred to as subsurface runoff (or runon). This part of the soil water budget was referred to as ground water runoff by Baver et al. (1972) and as subsurface storm flow by Whipkey (1965). In their seminal work, Dunne and Black (1970) used a shallow interceptor trench at the base of a slope to assess subsurface storm flow from a forested watershed. Only modest subsurface flow was permitted to collect in the drainpipe that was constructed in the bottom of the interceptor trench because overland runoff was diverted away from the trench. Although subterranean runoff has gotten far less attention than overland runoff in the literature, it can nevertheless be very substantial when poor hydraulic conductivity strata are present on sloped terrain. For instance, Hubbard and Sheridan (1983) found that subsurface flow accounted for roughly 99 percent of the nitrate in runoff from a field in the southeast of the United States.

Storage

Depending on the soil's geographic location and position in the landscape, significant temporal fluctuations in the amount of water it can hold may happen. Since the amount of water typically held in a soil profile is roughly the same at the start and end of the year of interest, this component is sometimes overlooked in estimations of the yearly soil water budget. A technique like TDR can be utilized precisely in the field to measure the initial and ultimate soil water storage components for shorter-term soil water budgets.

SOIL MOISTURE BUDGETS, IRRIGATION NEED

There is still debate regarding the relationship between actual and prospective transpiration. However, absolutely adequate water-budgeting techniques have been established that have proven to be of the highest practical usefulness in crop planning, irrigation scheduling, and water-resource management. These techniques were created to address specific conditions. For practical purposes, the amount of water in soil is typically described as "depth of water per indicated depth of soil." The following is a brief summary of the main working assumptions regarding the rate of water extraction:

- Between field capacity and the wilting point, water is equally available; or, evapotranspiration decreases linearly with increasing water tension (as proposed by Thornthwaite and Mather in 1955); or, a compromise is reached in which actual evapotranspiration initially proceeds at the potential rate before declining quickly in an exponential (or roughly equivalent) fashion. Choosing the transitional point still demands judgment.

practice in the humid temperate regime of the British Isles. It is assumed that:

- The root system is essentially limitless because there is just evaporation through the plants (transpiration), which occurs at the potential rate throughout.
- On April 1, the soil has reached field capacity (a reasonable assumption for the region).
- Any extra rain that falls above transpiration is utilized to replenish the soil to the field's capacity; any additional rain that falls is considered to have run off or drained away.
- The amount of water needed to restore the soil to its field capacity is known as the soil moisture deficit (SMD) (note: this emphasizes the importance of the concept when planning irrigation).

Table 15: *Simple water budget for dry summer in the United Kingdom*

	March	April	May	June	July	August	Septem-ber
RainfallR(mm)		53	20	91	29	14	25
Potential transpiration PT(mm)		51	84	79	97	83	58
R-PT		2	-64	12	-68	-69	-33
SMD(cum.)	0	0	64	51	119	188	221
Surplus		2	0	0	0	0	0

Source: Wieringa, J., & Lomas, J. (2001).

The scenario produces an exceptionally high cumulative deficit of 221 mm. It would require a crop stand with a robust root system capable of drawing water from such a soil block, as well as a homogeneous soil block of about a meter in depth, initially at field capacity. An established fruit orchard could serve as an example in the real world. In this scenario, water is assumed, and the model is treated as non-limiting. The table below illustrates a more realistic computation of the identical situation that takes into consideration the constraints placed on transpiration by the plant/soil interaction. It is presumptive that:

- The soil is at field capacity on 1 April;
- In response to an evaporative potential, the first x mm is drawn at the potential rate. When this supply is exhausted, the next y mm can only be released if a potential ny is -exerted. When (x + y) mm have been drawn, a further z mm requires a residual potential of mz to release it (where m>n);
- In this particular case, it is assumed that;

Stage I storage - the first 50 mm is drawn at the Potential rate;

Stage II storage- the next SO mm is drawn at half the potential rate (n = 2);

Stage III storage - the final 25 mm is drawn at quarter potential rate (m = 4);

- When R>PT, the surplus replenishes storage Stage I before affecting Stage II, water in Stage I being available for transpiration at the potential rate;
- Any surplus from Stage I is made available to Stage II and this water can be called upon at half rate; surplus from the second stage passes to Stage III and can be released at one-quarter potential rate.

Table 16: Specimen water budget for dry summer in the United

	March	April	May	June	July	August	September
R(mm)		53	20	91	29	14	25
PT (mm)		51	84	79	97	83	58
R-PT		2	-64	12	-68	-69	-33
SMD (cum.)	0	0	64	51	119	188	221
Storage (0-50)	50	50	0	13	0	0	0
Storage (51-100)	50	50	43	43	16	0	0

Storage (100-125)	25	25	25	25	25	16	8
Actual transpiration:							
From rain		53	20	78	29	14	25
From Stage I		-	50	-	13	0	0
From Stage II		-	-	-	27	16	0
From Stage III		-	7	-	0	9	8
Total actual transpira- tion		**53**	77	**78**	**69**	39	33

Source: Wieringa, J., & Lomas, J. (2001).

Following rounding off, there are slight variations in the quantities in the preceding tables. The process will be demonstrated by deriving the real transpiration for the months of July and August. If the July rainfall of 29 mm evaporates at its potential pace, there will be a requirement of 68 mm. In order to achieve this, Stage I storage can only provide 13 mm at full rate (thus lowering Stage I storage to zero), leaving a potential need equal to 68 - 13 = 55 mm. However, this only allows for the extraction of 55/2 = 27 mm from Stage II storage at half the possible rate. There are now only 16 mm (43 minus 27) in Stage II storage. Storage in Stage III is unaltered. Nothing from Stage I storage and 16 mm from Stage II storage can satisfy the 69 mm potential demand for August; but Stage II storage requires transpiration equivalent to 16 x 2 = 32 mm potential, leaving 69 - 32 = 3 7 mm to be supplied from Stage III storage. Only 1/4 x 37 = 9 mm of Stage III, where 25 - 9 = 16 mm is still present, can be released by the 37 mm potential. It should be emphasized that the actual transpiration totals are 348 mm, as opposed to the 452 mm total (see second line of table for PT). Numerous models have been developed to consider the following elements:

- The seasonal changes in the color, degree of ground cover and stage of growth of the crop;
- A plausible assumption regarding the accompanying changes in the size of the root mass is that the roots are steadily growing downward at a rate that allows them to collect water from increasingly deeper levels in order to fulfill and satisfy potential needs.

The daily estimations of potential evapotranspiration and observed data on precipitation are inputs into a method for estimating the amount of soil water that will be needed, known as a flexible budget (Baier and Robertson, 1966). Using physical and biological principles of water transport into the soil and water loss from various soil depths, this technique mimics change in daily soil moisture content and provides estimates of Actual Evapotranspiration (AET) and soil moisture levels in six zones of the soil profile. By comparing the soil moisture estimations to actual measurements taken in the field (Baier, 1969, 1973), as well as indirectly through correlations between these soil moisture estimates and wheat yields, the validity of the soil moisture calculations has been confirmed (Baier and Robertson, 1968).

CHAPTER

6 Climate, Representativity and Topography

CONTENTS

OVERVIEW

By measuring the prevalent weather factors whose periods of observation span over time, the sorts of climates that prevail over a region are identified. The portrayal of the actual climate is better the longer the observation time. Although the observed weather events span a wide variety of scales, they are all actually parts of a continuum and none are discrete. As a result, there is dispute regarding the scale limitations when attempts are made to categorize weather events into separate scales. The type of heat, moisture, and momentum exchange between the Earth's surface and the atmosphere determines the characteristics of the climatic spectrum. Everywhere on Earth, the exchange processes between the Earth's surface and atmosphere are governed by the same physical rules. There is debate regarding the areal scale in climatology, just as there is with the issue of meteorological scale. Two typical strategies are described below to demonstrate this. The first division is into three parts, while the second division also includes another category that is frequently found in regional meteorological studies.

THE BASIC SCALES

By having a network of observation stations spaced at intervals of a few square kilometers, about 1 km, and approximately 20 km, respectively, these features are revealed. The subdivision of the climatic phenomena provides a basis for dividing the climatology into micro-, meso-, and macro-climatology (Munn, 1966). .

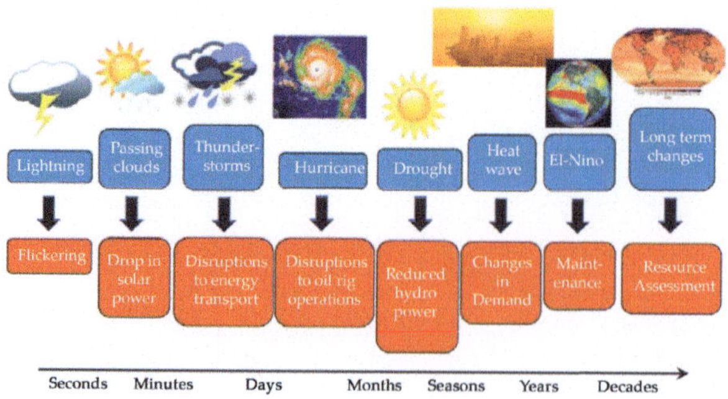

Figure 36: *Time and space of various atmospheric phenomena.*

Source: Božiček A, Franc B, Filipović-Grčić B. Early Warning Weather Hazard System for Power System Control. Energies. 2022; 15(6):2085. https://doi.org/10.3390/en15062085

The table below illustrates how the classification schemes use horizontal and vertical properties as criteria.

***Table 17**: Scales of meteorological motion systems*

	Approximate characteristic dimensions			
Motion system	Horizontal scale (km)	Vertical scale (km)	Time scale (n)	Total energy
Macroscale				
Planetary waves	5×10^3	10	5×10^2 to 2×10^3	-
Synoptic perturbations	5×10^2 to2×10^3	10	10^2	Average depression
Mesoscale phenomena	1 to10^2	1 to 10	1 to 10	Average thunderton
Microscale phenomena	$<10^{-1}$	$<10^{-2}$	$10^{-2}$to 10^{-1}	Average wind gust

Source: Barry (1970).

Microclimatology

Geiger (1966) highlighted the significance of the air layer below the established level of the atmosphere, which is approximately 2 m and deals with the "climate near the ground." The horizontal extent of the circulation at this level may only be a few tens of kilometers. Changes in both the vertical and horizontal planes can affect the magnitude of every meteorological component in this air layer.

While the horizontal fluctuations are caused by changes in the kind and moisture of the soil, variations in surface slopes, and by the type and height of vegetation cover on it, the vertical variations are caused by how close to the ground something is. Microclimate can also be described using confusing terms like spot-climate, position climate, and miniature climate. The microclimate, according to Huschke (1959), is a finely tuned climatic structure of the air space that extends from the Earth's surface to a height where the immediate impacts of the underlying surface cannot be differentiated from the overall local climate. According to Munn (1966), the surface boundary layer, which is the lowest hundred meters of the atmosphere, is the only part of the atmosphere where microclimatology may be conducted.

Meso-climatology

In meso-climatology, the configuration of the ground, the kind of soil, and the vegetation cover are seen as local elements that influence the climate that predominates in a certain location and is referred to as the local climate and are only subject to modest changes over time. This kind of phenomenon (mesoscale feature) has a horizontal extent of several hundred kilometers. Thornthwaite (1953) proposed the idea of topo-climate as a middle ground between macro- and micro-climates; however, Scaetta (1935) suggested changing the term to meso-climate, and Utaaker (1974) believed that the term topo-climate is better suited for local or meso-climates since the topography of the land is the primary causal factor behind local climatic peculiarities. According to Konstantinov et al. (1974), the underlying surface has an impact on how microclimate and local climate are formed. The mesoscale events were described by Barrett (1974) as "local embroideries upon a wider, more generalized climate image."

Table 18: *Scale of climate and its corresponding meteorological phenomena*

Climate	Horizontal distribution (m)	Vertical distribution(m)	Example of climatic phenomena	Lifetime of corresponding meteorological phenomena(s)
Micro-climate	10^{-2} - 10^{2}	10^{-2} - 10^{1}	Climate of greenhouse	10^{1}– 10^{1}
Local climate	10^{2} - 10^{4}	10^{-1}- 10^{1}	Thermal belt of slope	10^{1}– 10^{4}
Meso-climate	10^{3}-(2×10^{5})	10^{0}- (6×10^{3})	Climate of basin	10^{4}- 10^{5}
Macro-climate	(2×10^{5})- (5×10^{7})	10^{0}- 10^{5}	Climate zone, monsoon region	10^{5}– 10^{6}

Source: Yoshino (1975)

Macro-climatology

This relates to the major characteristics of the climates that are controlled by the large-scale atmospheric circulation systems across significant regions of the world or the majority of a continent. It includes taking global mean weather bands into account. A continent's macroclimatic features are defined by large-scale disturbances that span thousands of kilometers, including moving cyclones and anticyclones. Regional macro-climatology examines both the overall tendency of synoptic weather aspects as well as differences in the number, intensity, and seasonal distribution of climatic factors. Changes in the geographic distribution of a meteorological event are partially caused by the high relief.

Scale and Local Climates

Grasp the macro-climatology of a continent requires a thorough understanding of the interactions between climatic patterns and the Earth's surface as well as the impact of major atmospheric systems on even lower levels of the atmosphere. Yoshino (1975) believes that the term "local climate" is midway between microscales and mesoscales in a review of climatic scale (see Table above). He contends that meso-relief, vegetation masses, and other climatic elements govern local climate. A layer of air that may be seen at locations that are tens or even hundreds of meters high exhibits the features of the local climate. Furthermore, meso-climatic phenomena have been characterized as those that are independent types of local climate and are restricted to the size of objects that constitute single independent units in geographical taxonomy. In the USSR, there are two opposing viewpoints on scale: the first holds that local climate is a subclass of a microclimate in a larger sense, combining local climate and microclimate, and the second holds that local climate is nothing more than a subclass of a microclimate in the first position.

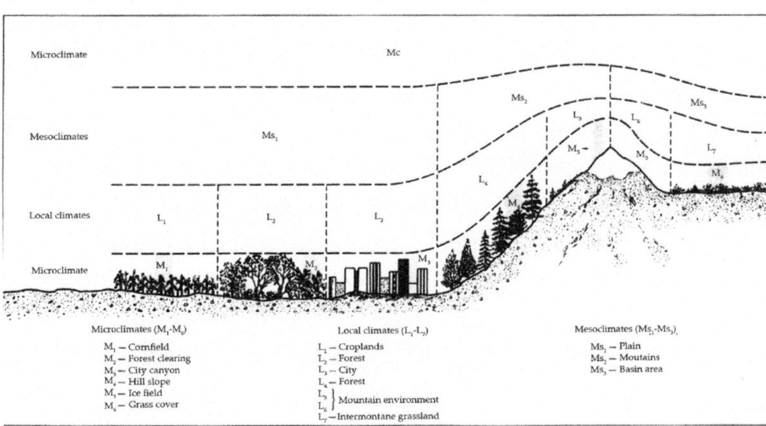

***Figure 37**: Area scales of climatic investigation.*

OBSERVATION REPRESENTATIVITY, EXPOSURE AND SAMPLING

The instrument is merely one link in the chain of meteorological observation, and its accuracy is especially important. These other links are significant ones:

- The exposure of the instrument, i.e., its siting and mounting, and surrounding barriers;
- The sampling procedure, i.e., duration and frequency of observations;
- The observation and recording procedure (including e.g., calibration).

The method of managing these links in the chain was optimized for typical meteorological stations in the 1950s. Since then, devices that are standardized and made to monitor the macroscale atmosphere have been made available to weather forecasters and computer climatologists. As a result, GM-II does not go into great detail about exposure or sampling because it is not necessary. You are operating properly if you adhere to the WMO's observational guidelines for predicting. However, because they operate at smaller scales, agricultural meteorologists cannot trust that this observation product would sufficiently meet their own information needs. Therefore, they must choose which exposure and what kind of sampling will adequately match their specific problem. They can only choose the right instruments after responding to these queries. Their synoptically-minded coworkers may inquire as to why their station measurements are insufficient if it subsequently appears that alternative hardware and observation techniques are needed for agro-networks. All of the observations acquired must be "representative," but this is a fundamental criterion for a standard station that is all too frequently misunderstood. When someone claims that representativity means that "at a given time of observation, observations of some meteorological element should be approximately the same at two or more observing stations judged to have the same exposure," he is not referring to representativity but rather to the accuracy of observation. In his case, the statement "judged to have the same exposure" conceals the question of representativity. The term "representativity" refers to the quality of an observation to offer a valid average value of the observed parameter in a predetermined space that we want to represent. This quality is dictated by the exposure, instrumentation (if any), and sampling technique. The atmospheric boundary layer in the area between 20 and 100 km from the station is typically the space to be represented for conventional network stations. This establishes that an average sampling approach across a time range of 10 to 1 hour is adequate. As a result, choosing the right instrumentation is rarely a pressing concern because even slow-responding meteorological devices can typically adjust to parameter changes within an hour.

We'll talk about the best exposure for each parameter below. The problem is less predetermined when it comes to agrometeorological surveillance. Some stations might be set up to offer climatological values in, say, a specific valley; in that scenario, the sampling and instrumentation standards for typical synoptic stations might apply. However, other stations are set up to reflect more intimate areas, such as particular orchards. Even though a Stevenson screen may perform exceptionally well at climatological stations, we cannot use it to detect an hour of frost in an orchard because, in light winds, a thermograph in the screen needs at least an hour to adapt to a temperature change (assuming no other differences are caused by other factors such as radiation error), and because the 1.5 m screen height is not ideal for detecting what is happening at ground level. The temperature on the screen is therefore not indicative of the orchard. This example demonstrates that representativity is not a fixed characteristic of a (well-executed) observation or a property that is universal like accuracy. It depends on the parameter's characteristics (such as radiation, precipitation, wind, etc.), as well as the use for which it is designed. Unless the circumstance for which an observation is representative is described, it makes no sense to simply term an observation "representative." Accurate measurement is not the same as representative observation, which necessitates appropriate and strategic placement of instruments. The location of the observation and sampling frequency are other crucial factors. It is conceivable to tailor the observations so closely to one application that it makes it impossible to use them for other purposes. A monthly precipitation sampling frequency, for instance, is valuable for climate atlases but meaningless to a hydrologist or meteorologist who does operational research.

Since user needs appear to differ greatly in terms of spatial scale, representativeness criteria are user-dependent. This raises the issue of how to configure a station network so that the majority of the key applications are adequately supplied. To focus on the sources rather than the applications is a general answer to the issue. Incoming solar radiation, both short-wave and long-wave, as the primary source of heat, large-scale wind, say at geostrophic level, as the primary source of motion, and precipitation, as the source of water, are the three main source parameters that have an origin high up in the atmosphere. It is obvious that having a horizon that is as free as feasible is important for "representative" radiation measurement. Obstacles in the area shouldn't stick out more than a few degrees above the horizon, and they shouldn't face the Sun (e.g., South in the Northern Hemisphere). As a result, radiation or sunshine meters ought to be installed at the highest point that is nevertheless conveniently accessible for cleaning the dome or exchanging cards (Campbell-Stokes, Robitzsch) (of a radiation meter). The large-scale airflow is the source of wind in the surface layer, far above any surface roughness or obstructions that would slow it down. Anemometers should be situated quite high, but not on a building, because such obstructions alter airflow considerably more than most people are aware of. This is important if one wishes to obtain wind data that are most relevant for various applications, both local and remote. One of the few requirements that is a strict necessity and not simply a suggestion is the WMO's recommendation for a 10 m mast in an open area. In most soil conditions, climbable 10 m masts can still be built very simply even without man wires. An important source of operational inaccuracy is the fact that FAO only uses wind measured at a height of 2 m (Allen, et al., 1994). We want to know how much precipitation actually falls to the earth, but measuring precipitation is difficult because it is so wind-dependent. Once more, the WMO suggests a compromise: measurement at the lowest level physically possible, provision of a wind shelter, but the absence of obstructions of height H within a radius of 4 H. (in order to prevent obstacle drip into the rain gauge).

WIND BEHAVIOR

Now is the time to think about air movement and the lower atmosphere's thermodynamic characteristics. The observation height of the wind is frequently not even stated in evaporation procedures, much less debated. However, as wind plays a significant role in the transportation of heat, water vapor, and CO_2, having some understanding of how it behaves on actual terrain (rather than just over idealized expanses of endless grass) should not be viewed as an extravagance.

The behavior of airflow around surface obstructions, particularly around shelterbelts, is the foundation for this study of wind. These play a significant role in agrometeorology, and a basic examination of how they work naturally leads to a more comprehensive discussion of how surface features affect wind behavior.

Wind around Barriers of Varying Porosity

Crops are not always harmed by wind, and in some cases, it is even necessary. However, strong winds have the potential to damage property and erode soil. Therefore, the deployment of shelterbelts and windbreaks frequently reduces the destructive effect of severe winds across a certain area. A single or numerous rows of forest trees, or a block of trees, with or without bushes below, constitute a shelterbelt. The phrase "windbreak" typically refers to a shorter hedge or to a man-made shelter, such as a wall or fence constructed of non-vegetable materials. All kinds of wind obstructions are referred to as barriers, but those with straightforward geometrical shapes are especially so. Here, the terms "barrier," "windbreak," and "(shelter)belt" are used synonymously. All linear dimensions (i.e., crosswind length of a barrier, downwind distances from a barrier, and vertical lengths above ground level) should be expressed in multiples of the barrier's height (H) when comparing the impacts of different sizes and types of barriers. A barrier's percentage porosity or permeability, which is the ratio of the open area within the barrier to the total barrier area, where both of these areas are considered perpendicular to the barrier axis, must also be specified. Typically, wind speeds within the barrier's zone of influence are stated as a percentage of the 'free' (i.e., unhindered) wind speed.

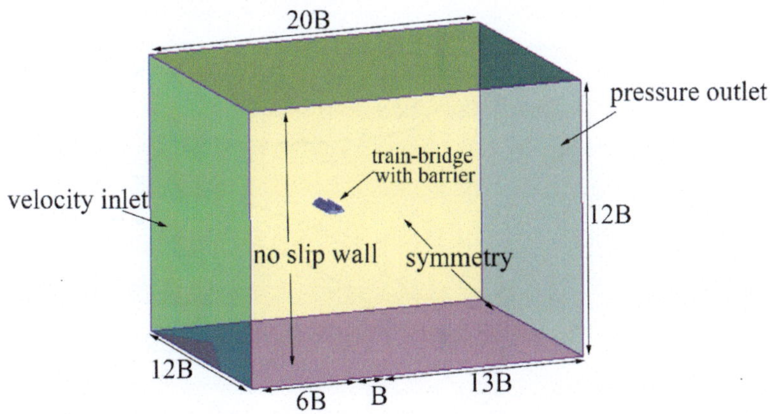

***Figure 38**: Airflow as influenced by a solid barrier placed normal to the flow.*

Source: Lou P, Tao W, Cai C, He X, Zou Y, Ai Y. Influence of Wind Barriers with Different Curvatures on Crosswind Aerodynamic Characteristics of a Train-Bridge System. Applied Sciences. 2022; 12(3):1747. https://doi.org/10.3390/app12031747).

However, not all publications explicitly state whether their "free" reference speed is at the same height as the location in the barrier influence zone being studied or at a conventional reference height. The barrier will be assumed to have infinite length along its axis in the basic explanation presented below, making a two-dimensional flow description sufficient.

Since the air closer to the barrier is obstructed in its path and consequently experiences pressure increase, the flow at a distance of 3 to 10 H is the first to recognize its presence upwind. Stalling air at pressures below H creates a cushion-like structure just in front of the barrier, frequently accompanied by a "bolster eddy". The flow then ascends over this into the aforementioned displacement zone. In this zone, the layer immediately above the top height of the barrier must be traversed by a sizable flow of air rising from lower levels. Therefore, wind speeds are relatively high immediately above the barrier and for a short distance downwind in the displacement zone, resulting in a "jet" of accelerated air. The accelerated air is able to flow downward after passing the barrier, but due to inertia, it continues to move mostly forward. Behind the barrier, a cavity zone with low speed and pressure forms below this jet, and its lower pressure pulls the jet downward. Large lee eddies arise in the cavity zone and can reach a height of 1.5 to 2 H when they are 3 to 5 H downwind from the barrier edge.

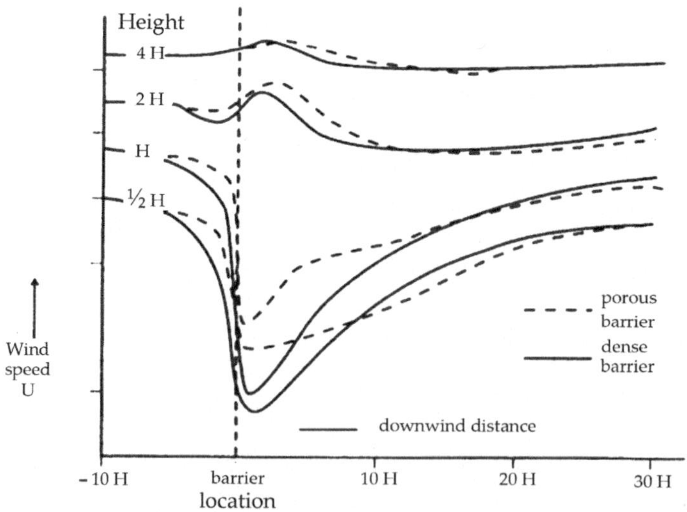

***Figure 39**: Actual wind speeds at various heights before and behind a barrier of height H.*

Source: Nageli, (1943).

The air is quite turbulent at ground level at this distance, as shown by crop damage and surface wind erosion. Depending on atmospheric stability, the cavity zone's downwind ranges from 6 to 12 h. (Counihan, et al., 1974; Seginer, 1975a). Because it must incorporate increased eddying near the belt, a local decrease in mean wind, and recirculation of unrefreshed air, modeling of evaporation in the cavity zone generated by a solid barrier is challenging. The turbulent cavity flow and the displaced jet-flow gradually combine into the barrier wake, where the lee eddy turbulence is broken up into smaller eddies. This wake, which has wind speeds that are substantially lower and more turbulent, stretches 20 to 30 H downwind of the barrier and can occasionally be seen up to 50 H (Jacobs, 1985).

According to field observations, the flow above about 3 H is not significantly disrupted, with the possible exception of the area immediately above or within 1 or 2 H downwind of the barrier. There are two significant changes in the airflow over a solid barrier if we now replace it with one that is permeable (let's suppose it has a 50% porosity). First, because some airflow can now travel through the barrier, the upwind stagnation is reduced. Due to this, lee eddies and wind speed reductions in the cavity zone are frequently significantly less. Second, depending on porosity, the wake now extends farther downwind, and velocity reductions are readily apparent to a distance of 30 H. This is because the cavity's pressure is no longer extremely low, which prevents the displaced jet-flow from being dragged downward and causes a slower restoration of the surface speed. The figure above shows the extent and resultant degree of wind speed reductions behind a two-dimensional shelterbelt for both a dense and porous barrier. The following characteristics stand out: speed reduction upwind of the belt, acceleration above it, maximum retardation right behind the band dependent on porosity, wake extending to 20-30 H downwind, and negligible effect at heights over 3 H. The illustration above demonstrates that choosing the right porosity for the best wind reduction involves making trade-offs. A porous barrier provides less reduction but it lasts longer. Medium-dense barriers with a permeability of roughly 35 to 55 percent tend to produce the best results for major wind protection over vast regions.

While accurate methods for measuring the permeability of natural shelterbelts of trees with or without undergrowth and heavy shrubs, etc. do not yet exist, good estimates of permeability are available for an artificial barrier of, instance, bamboo fences or paling, or stone walls (Bean, et al., 1975). Additionally, the thickness of the belt is greater in an oblique wind than in a wind that is coming at the belt perpendicularly. It is necessary to take into account both the overall screen density (assessed visually) and the size of the individual holes (referred to as the mesh size) when determining a belt's porosity or permeability for airflow (dynamic porosity). The shelter effect increases with decreasing mesh size for a given optical density. The material's characteristics matter when the meshes are smaller than a few millimeters in size. Natural belts made of multiple rows of twigs, leaves, and branches have a higher effective density than would be perceived optically when compared to the flat surface of wire or board. Additionally, the flow resistance of the hole increases if the thickness of the screen is greater than a few times the diameter of the hole (Gandemer and Guyot, 1981). Up until this point, we had only talked about barriers with uniform porosity. However, a barrier may be more open towards the bottom (e.g., a row of broad-crowned trees devoid of undergrowth) or higher up (e.g., more open) (scrawny trees and much undergrowth). Although considerable porosity near the ground does produce "jet" effects below the belt, it typically results in a wind reduction pattern that extends farther downwind.

Wind Reduction by Shelterbelt Arrays

Shelterbelts may serve a variety of functions. They may simply serve to moderate the overall windiness close to the ground, they may have been designed to withstand sporadic, extremely strong gusts from that direction, or they may be there to prevent sand, salt, or snow from encroaching. Real shelterbelts are only so long before air rushes around the end and encroaches from the side into the protected area, degrading the sheltered zone. This happens when a persistent wind blows perpendicular to the belt.

This erosion limits the maximum extent of the protected area that its height and structure may theoretically achieve, even close to the center of the belt, for a belt with a limited crosswind length (i.e., perpendicular-to-the-wind). To protect a useable area despite lateral erosion and wind direction changes, a belt's crosswind length must be at least 12 times its height. The sheltered area has an approximately parabolic form when the crosswind is long enough.

The illustration below shows some of the characteristics of such a region in the event of an impermeable and a 54% porous shelterbelt. It should be noted that the impermeable belt provides effective protection (a 3x reduction in average wind speed near the belt) but also suffers from relatively significant lateral encroachment that leads to erosion of the protected area. This is because the flow around the ends of this solid barrier is quite strong.

The porous belt, on the other hand, reduces wind speed somewhat over a region that is roughly three times larger than the area protected by the impermeable belt.

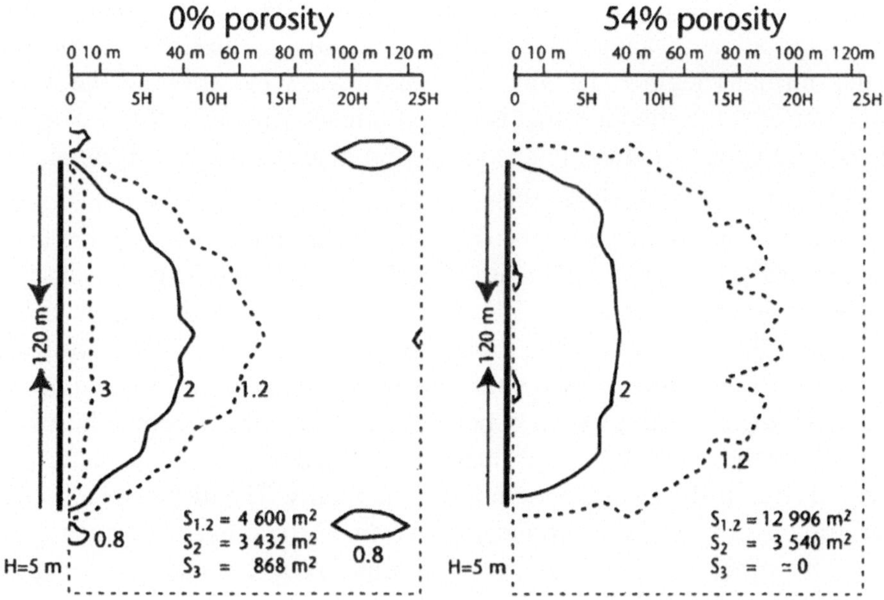

***Figure 40**: Isolines of protection factor.*

Source: Gandemer and Guyot, (1981).

Because there is only one impermeable belt, the high-speed flow around its ends is not as powerful as it could be, which allows extra flow to disperse sideways. However, the wind will tend to blow through the gaps in a succession of belts with spaces between them for the passage of cattle and machines in strong "jets." Trees or bushes on either side of the gap are particularly vulnerable to wind damage, and such gaps are often the origin of following progressive blows in forests. The risks from "jets" cannot be completely removed, but they can be reduced by protecting the gap's sides with hardy shrubs or wire mesh, or, if the circumstances allow, stagger the shelter belts to position the gaps in a zigzag pattern. Belts of the "alley" kind, or rows of tall trees without branches or plants as an under-storey, exhibit a similar "jet" phenomena. These belts typically have a single or double row of trees alongside roads and highways. As a result of airflow passing through the bottom portion of the belt due to gaps between the tree trunks, this sort of belt's surface wind speed may even increase below the tree canopy. The limit of the zone protected by alley belts extends significantly further downwind, despite the fact that the maximum wind reduction at distances of roughly 5–6 H is, of course, somewhat less than behind more homogeneous types of belts. The overall size of the downwind covered area, when the wind comes at a shelterbelt at an angle, is roughly proportional to the cosine of the angle to the normal (Seginer, 1975b). Thus, until the wind angle to the normal surpasses 30°, the shielded area is not significantly reduced. Even at greater angles, the general parabolic shape is maintained; however, if the angle is more than 45°, there is little cover.

The influence of the thickness (or distance through) of a single shelterbelt has primarily been researched in regard to broad, forest-type belts with cross-axis diameters of at least 5 H. The main conclusion is that compared to a similarly thick but narrow belt, the sheltering zone beyond the downwind forest border does not extend as far. The main air stream is directed upward at the windward side of such a belt, but it gradually becomes horizontal flow across the extremely rough surface provided by the treetop canopy and eventually exits the lee edge traveling horizontally. The displaced air stream then travels closer to the ground relative to a single barrier than it would have done otherwise. The features of an impermeable barrier are effectively acquired by a very thick band, and the protection distance downwind is correspondingly reduced. The shelter effect between belts tends to be greater in relation to a succession of parallel belts than it is in relation to the lee of a single barrier. According to flow studies (such as Beljaars, 1982), belts with spacing of around 100H can still generate almost twice as much shear stress on large-scale airflow as a treeless yet vegetated plain.

These impacts on the wind environment over a greater area are equally important with vast interspaces. However, downwind wakes combine with upwind shelter zones of the following belt when the belts are fewer than approximately 20 H apart in the wind direction. This offers wind protection for the entire area and has implications for wind profile modeling. The efficiency of the shelterbelt spacing must be taken into consideration. Because the ideal separation distance depends on the wind conditions and other variables, there is no universally optimal separation distance for rows of parallel bands. The belts can be significantly farther apart than when the same overall frequency of strong winds is dispersed over a number of directions when damaging winds originate primarily from one direction or from two directions that are directly opposed to one another. The intended amount of the average wind reduction across the entire area that is to be protected determines how far apart shelterbelts should be placed. Other considerations include the need to shade the ground from direct sunlight, variations in soil moisture and air temperature, and the kinds of plants and crops that may need protection. When the efficient use of farm machinery necessitates large-sized fields, the space between belts also needs to be substantial. When the goal is to reduce the mean wind speed generally over a large region as opposed to when a wind threshold value—such as an 8 m s-1 threshold value for soil erosion—must be excluded at all times and at all locations, the distance between belts can typically be higher. On broad plains, the most common system of shelterbelts consists of parallel belts that are positioned perpendicular to the main local wind direction and generally near together. These belts are supplemented by other belts that are more widely spaced and parallel to the main wind direction. There are recommendations for a network of parallel and perpendicular belts in windy regions of Germany that are spaced 250-300 and 500-600 m apart, ten meters high, and cover an enclosed area of around 0.15 km2. Systems of forest belts that encompass typical plots of 400 x 1 200 m2 are advised for the southern regions of the Russian plain, particularly in Ukraine. In order to reduce wind speed generally, shelterbelts are planted at 30–50 H separation distances in some parts of Western Europe, such as the Netherlands.

There have been very few studies on specific uses of shelterbelts in mountainous areas (Rutter, 1968). However, it is clear that their placement must take into account their potential impact on the formation or absence of nighttime cold katabatic air pools. In unique situations, damaging winds can only travel in a small range of directions. These situations are linked to valley winds that must blow in one direction (or two directions that are diametrically opposite), or, on a larger scale, specific regional winds. For protection against the frequent, extremely powerful, and extremely cold dry north wind known as the Mistral, a particularly dense network of parallel belts (at roughly 4 to 6 H apart) is required in the horticultural area of the Rhone Valley in southern France. The airflow fundamentally changes from the scenarios covered above at shelterbelt separation distances of less than 3 H. Perry, et al(1969) .'s wind-tunnel investigations of the airflow over a surface with transverse rectangular trenches of the same depth but spaced differently serve as an example of this. There were two distinct forms of flow. In general, qualitative terms, it was discovered that independent circulations of roller eddies were created within the cavities for D/H1 (D = distance between objects), only loosely associated with the flow above. The upper flow was mostly connected to an effective base plane immediately below the tops of the trenches, at a fraction of the obstacle height. This flow was referred to as "d-type" by Perry, et al. (1969). This mimics the situation of a dense forest or crop in nature, with the flow "skimming" over the tops of the vegetation.

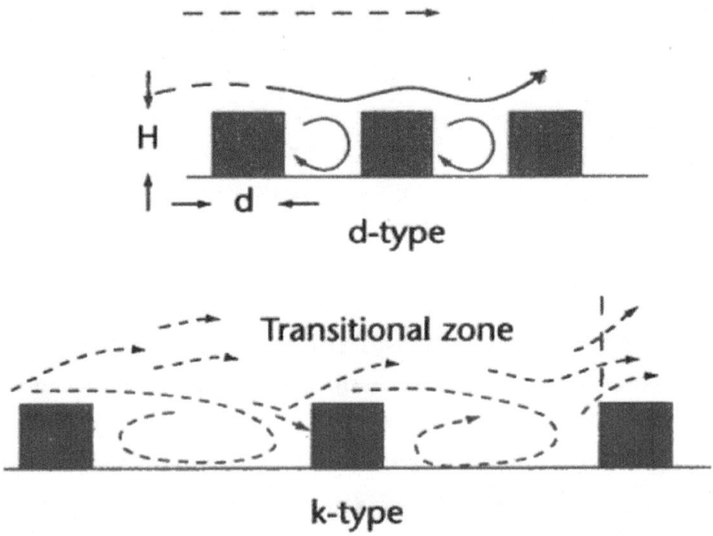

Figure 41: *Two-dimensional surface flow patterns.*

Source: Perry, Schofield and Joubert, (1969).

As D/H increases, the external wind begins to penetrate the gap. When D/H ≈3 or more, the externalflow circulates in the cavity, and the edges generate a turbulent wake which is fedinto the main stream. The result is the formation of a transitional zone, above which the steady external flow is established. This flow was referred to as "k-type" by Perry, et al. (1969). Since then, it has been demonstrated that these transitional zones, also known as "roughness sublayers," occur over d-type skimming flows in nature (Raupach and Thom, 1981). The zone height depends on interspace distance D, but reaches at least to 0.5H above the tops (Wieringa, 1993).

Anyhow, it is obvious that common shelterbelt-type airflow cannot establish itself, unless there is at least space for a cavity zone, say if D≈10H. By creating frictional drag, the collection of "roughness elements"—tree plantations of various sizes, from bushes to forests, tree belts, and hedges—covering large areas of the countryside tend to slow the wind. Based on a comparison of winds across two Jutland transects, one with significant surface roughness and the other with very few natural impediments, Jensen

(1954) reports such effects. When major changes in land use are being considered, these issues become more crucial. Even though it cannot be demonstrated that the removal of a single hedge, belt, or forest block will change the meso-climate of a region, the destruction of these features on a large scale can result in an important increase in wind and possibly a decline in the area's ecological quality, accelerating existing erosion processes, and unquestionably significantly reducing the variety and number of different habitats.

Wind Representativity at Topo-scale

Common terrain is typically full of barriers like shrubs, single trees, plots with significant vegetation, human constructions, and possibly hillocks or ditches, even if it is flat and there are no shelterbelts. These all alter the wind in a similar way to how shelterbelts do. In the wind-field, a change in surface roughness will be visible. If a fallow flat area is followed by crop or meadow vegetation, wind turbulence and its rate of height increase will gradually adapt to the new surface.

An Internal Boundary Layer (IBL; see Garratt, 1990) is the layer immediately adjacent to the new surface where wind structural adjustment has already occurred. The height of the IBL increases as the wind moves downwind across the new surface. The wind-field structure above the IBL is still influenced by surface conditions from the prior stretch of surface. It goes without saying that describing the wind-field very near to actual terrain surfaces is highly challenging. At heights exceeding 3 H, shelterbelt flow, on the other hand, behaves quite consistently. We can define an average flow speed U at that level or higher, which does not significantly change over horizontal lengths of 30 to 50 H. The average wind speed at levels above 3 H is discovered to grow with height z in the following manner across terrain with obstacles of height H:

$$U_2/U_1 = ln(z_2/z_0)/ln(z_1/z_0)$$

where U_x is the wind speed at height Z_x. In this logarithmic wind profile formula, the so-called roughness length z_0 depends on the degree of irregularity of the surface, and an unreliable approximation gives $z_0 \approx 0.15$ H. The logarithmic profile is valid for neutrally stable conditions, i.e., a dry-adiabatic lapse rate, and then it can be used at heights above 20 z_0,i.e., for z >3 H, up to about 100 m (Tennekes, 1973). Above 3 H, turbulence caused by the surface roughness does also conform to simple laws. If σ_U is the standard deviation of wind speed samples of 3 seconds duration (call them gusts) around a wind speed U averaged over 10 minutes, we find that:

$$\sigma_U = U_z/ln(z/z_0)$$

if σ_U and U are measured at the same height z. Moreover, a good estimate for u_{mx}, the largest gust with duration t at height z within those 10 minutes is:

$$u_{mx} \approx U + E_\lambda \, \sigma_U = U \ (1 + E_\lambda/ln(z/z_0))$$

where $E_\lambda \approx 1.42 + 0.3 \ \text{In} \ ((10^3/Ut) - 4)$ is the statistical ratio of the largest probable extreme and the standard deviation; e.g., if U≈10 m s^{-1}, $E_\lambda \approx 2.44$ for gusts of 3s duration. With averaging periods of an hour instead of 10 minutes, the probability of strong gusts increases: $(u_{mx}/U)_{60 \ min} \approx 1.10 \ (u_{mx}/U)_{10 \ min}$. For finding the most probable duration of the largest gust measured by a given combination of anemometer and wind speed recorder, information is given by Wieringa.Not all terrains - in fact, very few - are homogeneously covered with obstacles of a single height. If we have a field with crops of 0.3m height, and sayz_0 =0.05 m, and if we have measured U at some height z =>1 m, we can now estimate average wind and turbulence at other heights which are not less than 1 m. But suppose that there is a large hedge or a tree at the upwind side of the cropped field, how does z_0 change? The question indicates that z_0 shows a scale-dependence. If we have a field of mown grass, then $z_0 \approx 0.01$ m only for the average wind profile within the height of the IBL fully adapted to the grass - and this IBL-height is about 1 percent of the fetch, i.e., the distance to the upwind edge of the (grass) roughness. If we want to obtain U at higher levels, or to estimate turbulence, we have to take into account aspects of the (rougher?) terrain further upwind. Before calculating the change in z_0, we must extend the problem.

The larger-scale airflow, which is driven by pressure differences and possibly modified by large-scale topography, is the source of wind at the surface. It typically has a fairly stable value over a period of an hour and over a distance of several kilometers. If the terrain is covered with obstacles with maximum heightH_{mx}, the lowest height at which we can expect the occurrence of the larger-scale source wind is3 H_{mx} ·Assuming that for trees, or buildings, H_{mx} is not larger than 20 to 30 m, we can define a blending height z_b at 60 to 100 m, where the influences of separate surface features on the larger-scale wind field are no longer recognizable individually, but where the average wind is still well correlated with average wind observed near the surface. The association with surface wind drastically declines for heights more than 100 m for a variety of reasons (Wieringa, 1996). Since the wind at blending height is very uniform at topo-scale, it can be considered a source wind. However, using the logarithmic wind profile, we can still link the wind at heights between 2 and 10 m quite consistently.

The determination of a suitable roughness length for the relationship between, say, 10 m and 60 m height, is the practical challenge. The appropriate z_0 -value is one, which includes the integrated effect of all roughness variations in the upwind terrain near enough to influence the profile. A good estimate for this upwind fetch is $30z_b$, i.e., 2 to 3 km, and we must find the overall effective roughness for an upwind area of that size. Masts of at least 50 m in height would be required in order to calculate effective roughness from profile data, which are extremely unusual. However, as turbulence information preserves information on the roughness of previously traversed terrain over a significant distance, an effective z0 can also be ascertained from that source. The formulas given above for both cru and u_{mx} can be inverted to determine a z_0 –value from measured gustiness, and this proves indeed to be the correct effective z_0(Beljaars, 1987a).

***Table 19**: Davenport effective roughness classification*

Class		Roughness length (m)	Landscape description
Num-ber	**Name**		
1	Sea	0.0002	Open water, tidal flat, snow, with free fetch> 3 km.
2	Smooth	0.005	Featureless land with negligible cover or ice.
3	Open	0.03	Flat terrain with grass or very low vegetation, and widely separated low obstacles; airport runway.
4	Rough-ly open	0.10	Cultivated area, low crops, occasional obstacles separated by more than 20 obstacle heights H.
5	Rough	0.25	Open landscape, crops of varying height, scat-tered shelterbelts etc., separation distance≈15 H.
6	Very rough	0.5	Heavily used landscape with open spaces ≈10 H; bushes, low orchards, young dense forest.
7	Closed	1.0	Full obstacle coverage with open spaces ≈H, e.g., mature forests, low-rise built-up areas.
8	Chaotic	>=2	Irregular distribution of very large elements: city Centre, big forest with large clearings.

Source: Wieringa, (1996).

With this effective z_0 we can transform wind speed data between two nearby locations. When at an observation station we have measured a wind speed Us at a height Z_s, and the station upwind roughness is estimated to be z_{0s}, we can obtain an estimate of the wind speed at height z_b. Subsequently, we use the obtained U (z_b) to estimate the wind U_r at some nearby reference location with height Z_r, where z_{or} is the local upwind roughness. Taking 60 m as a working value for z_b, the upward-downward transformation formula is:

$$U_r = U_s \frac{ln(60/z_{0s})ln(z_r/z_{0r})}{ln(z_s/z_{0s})ln(60/z_{0r})}$$

Although this transformation is based on the logarithmic wind profile, which makes the assumption that neutral stability exists, it has been demonstrated in practice that it also works in scenarios where stratification is either unstable or stable (Wieringa, 1980).

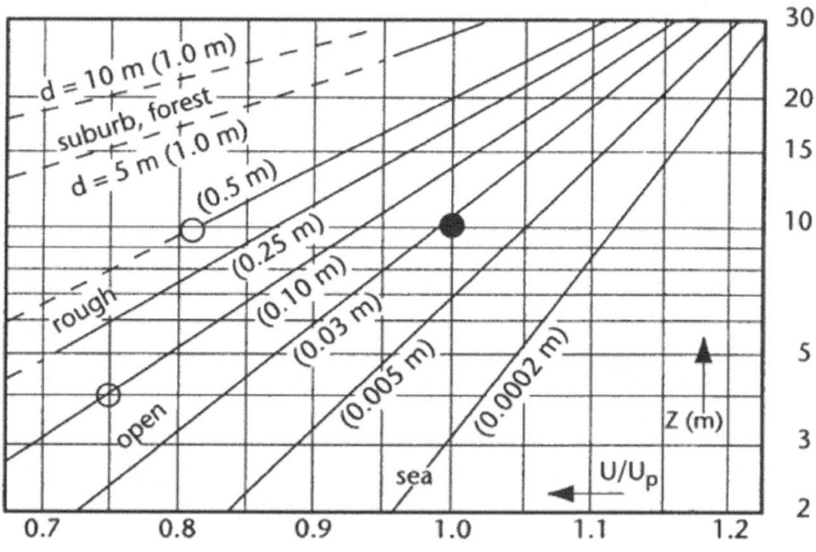

Figure 42: Transformation diagram for observed average wind speeds U.

Source: (Wieringa, 1986).

The reason is that any stability-caused deviations in the upward transformation to z_b are compensated by similar but opposite deviations in the downward transformation. However, for such compensation, the transformation's height intervals must be comparable. Therefore, the transformation nomogram, the figure below is restricted to heights Z_s and

Z_r which are=<30 m. In order to simplify wind transformation between two places with given height and upwind roughness, the nomogram below uses the transformation formula mentioned above. The Centre point is the observation situation officially required by WMO, namely 10 m over open terrain, for which $z_0 = 0.03$ m (roughness-class3). This WMO-reference wind is generally called 'potential' wind U_P. It is the wind speed that would have been observed locally, if the upwind terrain had been open and measurement had been made at 10m height. This terrain normalized U_P-value is a good parameter for large-scale wind climate studies because it has been corrected for purely local effects (e.g., the presence of the observer's house).

For example, if we assume that our station is located in 'roughly open' location, and that it observes at a 4 m height a wind speed $U_s = 6$ m s^{-1}; the appropriate U_P-value is 8 m s^{-1} (6/0.75), and at a nearby 'very rough' location we will observe at 10 m height 6 x 0.82/0.75≈6.6 m s-1. In the upper left-hand corner of the nomogram, the roughness isoline of 'suburb, forest' (roughness-class 7) is displaced. The base plane level, called 'displacement height d, typically lies at 2/3 H (Brutsaert,1975). Then:

$$\frac{U_2}{U_1} = \frac{ln[(z_2 - d)/z_0]}{ln[(z_1 - d)/z_0]}$$

The logarithmic wind profile is now in its modified version. Such a profile description is suitable for expressing wind above forests, and occasionally it is also suitable for wind above high crops. Now we must return to the question of scale-dependent differences in z_0. As a result of the need to take into consideration the wind structure of the entire 60 m high surface layer, including all surface inhomogeneities, the figures reported above are higher than those often seen handbooks. For wind analysis up to some height z over really homogeneous terrain, where over an upwind distance of at least 100 z the same roughness is present, smaller z_0-values may be appropriate.

For several agriculturally significant surface types, appropriate values of these are shown above; this list has been updated with all trustworthy reported field investigations conducted between 1965 and 1990. It is therefore far more correct than the z_0-list published by Sellers in 1965. Until now, most handbooks have copied Sellers' z_0-list, which is identifiable. by an unrealistically high roughness value for fir forest, namely a z-value of 2.83 m (derived erroneously by assigning to a forest of 5.5 m height a displacement height of only 0.1 m - d should be about 70 percent of the vegetation height). The Davenport classification of z0 -values are more appropriate when the situation is not truly homogeneous, such as when the observation field is tiny. Additionally, it is not recommended to translate wind data to other places using these homogenous z_0 -values.

Table 20: *Roughness lengths of homogeneous surfaces*

Surface type	Roughness length (m)
Concrete, flat desert, mudflat	0.0002-0.0005
Flat snow field	0.0001-0.0007
Fallow unplowed ground	0.001-0.004
Short grass and moss	0.008-0.03
Long grass and heather	0.02-0.06
Low mature agricultural crops	0.04-0.09
High mature. crops, e.g., grain	0.12-0.18
Continuous bushland	0.35-0.45
Mature continuous pine forest	0.8-1.6
Tropical forest	1.7-2.3

Source: (Wieringa, 1993).

TOPO-SCALE REPRESENTATIVITY OF METEOROLOGICAL OBSERVATIONS

Representative weather data are required for investigation or prediction of the interaction between the atmosphere and the plants. In actuality, this refers to topo-scale conditions such as temperature, humidity, precipitation, wind, radiation, etc., directly above and beside the crop or close to cowsheds. In the crop or the cowshed, information is ultimately needed at the microscale, but each plant situation's microscale questions must be addressed with the active participation of users. The techniques that meteorologists can use to generate topo-scale data include:

- Measuring parameters locally;
- Using measurements made elsewhere, near or far; and
- Supplementing measured or climatological parameter values by physical models.

For the sake of a straightforward argument, let's assume that the terrain is mostly level and possibly undulating with a few small, moderate slopes. On the subject of measurement, we can be succinct by saying that it essentially boils down to the availability of resources like time, money, and different types of labor. With the development of automatic weather stations, some authorities believe that the latter investment can be avoided, however, that belief is only partially accurate. To ensure the continuity that all weather observations require, automatic weather stations, unlike traditional weather observation methods, require highly qualified technicians to be accessible at short notice (i.e., within a day, not within a month). Depending on local conditions, either the high-tech approach or the use of less technically proficient but devoted observers at basic stations is preferred. Second, even with automated measurement, there are still a number of observations that require routine maintenance, most notably the monitoring of radiation and humidity. Automatic registration of a dirty pyranometer's output is not only meaningless, but it is also unjust to the uninformed data user. It is more cost-effective than many users believe to use measurements from other sources, such as official weather stations at the local or national level. Particularly operational users, like consultants or farmers, underestimate the work required to monitor meteorological data with a set minimum standard of quality.

There is a perception among the general public that weather monitoring is inexpensive because sunrise and rain do not, in and of themselves, imply expenditures for the user. This frequently shows itself as a pronounced refusal to cover the expenditures directly or centrally (at the state level). As a result, there are frequently not enough weather stations and not enough funding for them to operate effectively. This global issue necessitates that competent station data consumers verify the accuracy of the station data they obtain, maybe by paying a visit to the station. It may be necessary to make some changes when using meteorological data that was collected at a distance, whether it be five or SO kilometers, to guarantee the data are reliable for the intended use. Making brief "probe" observations at useful places for research objectives, followed by deriving adjustments in the form of linear mathematical regressions between "probe" observations and concurrent station data (e.g., Pedro and Gillespie, 1982). When true relationships are linear, this will work, but it must be repeated for every project, location, and season.

Operational users either use the data "as measured" or apply a standard rule-of-thumb correction (for example, topographical height corrections for radiation or "power-law" corrections for wind speed). In most cases, operational users are not interested in or able to set up such an experimental comparison. The highest climatological comparability, rather than just the shortest distance, should be the criterion for selecting a weather station as a data source, particularly in this operational instance. For instance, it may be preferable (depending on the goal) to use data from a more distant station that is also slightly inland than from a nearby beach station when meteorological information is required at a site some kilometers from the sea. The third option involves estimating the necessary parameter(s) using models that make use of data, physics, and location characteristics that are of a rather general nature, such as integral surface qualities.

Three characteristics (radiation, wind, and precipitation) may be reliably modeled on this basis at reasonably large scales because their main sources are atmospheric. The physical foundation for radiation estimation is based on the balance of three radiation components: net short-wave incoming energy, Kt (1 - a), downward atmospheric long-wave radiation, Lt, and outgoing terrestrial long-wave radiation, Li. There, it was demonstrated that estimating the long-wave components of the radiation balance is more practical than estimating the short-wave radiation from the amount of cloud cover or sunshine. Therefore, it would be ideal if more worldwide radiation observations were available. Topo-scale modeling is possible with information from wind speed stations. This wind speed can be translated to the desired application site if measurements are made locally at a well-exposed site at a sufficient height, or if accurate wind information is available, for example, from weather maps.

For this, it is important to understand the upwind (azimuth-dependent) effective roughness at the station and the application location. Then, a synoptic-scale network that is well-equipped is quite helpful. The main method for estimating precipitation at an application site is to interpolate between nearby stations by mathematically balancing their data based on distance. Some of the physical characteristics of rainfall are being used gradually. According to Gandin (1970), "optimal" interpolation, which relates weights to the statistical characteristics of occurring precipitation, did not appear to increase dependability as much as was anticipated (Kruizinga and Yperlaan, 1978). However, recent interpolation models now take into consideration the notable relative increase in precipitation at wind-sides of slopes (e.g., Benichou and Le Breton, 1987). Radar can also provide effective operational support for the interpolation of precipitation (Kessler and Wilk, 1968; Wilson and Brandes, 1979; Collier, 1986), and radar coverage is progressively expanding. Therefore, physically-based translation or estimation approaches to make the best use of the station networks already in place are becoming available for radiation, wind, and to some extent precipitation. Similar in structure, the models take into account surface factors that are integrated at topo-scale for larger-scale atmospheric sources such as global radiation, wind at blending height, and rain-bearing clouds. For straightforward geographic scenarios, the end result is:

- With the right handling, information on radiation and wind speed can be reliably extrapolated from a station over topo- and meso-scale distances;
- Interpolation between stations for precipitation is becoming increasingly accurate for rain brought on by fronts and other significant weather systems. Interpolation or extrapolation of station measurements of precipitation will still need a dense observation net due to the geographical variability of showers, and even then, conclusions can only be operationally reliable at local scales if they are supplemented by regular radar observations.

Other weather metrics' primary sources are found at or close to the surface. It is basically impossible to address their topo-scale behavior in the same way because of this feature, which necessitates modeling in conjunction with Geographic Information Systems (GIS). When there is no condensation (fog, dew), or evaporation, the air humidity, which is the atmospheric density of water vapor, is generally stable over short distances (advection over wet surface).

Dewpoint is also a rather constant humidity parameter over some distance in space or time. Relative humidity (R.H.) is definitely not spatially constant, and R.H. data should always be transformed into P_v or T_d with the use of the simultaneous station temperature before any attempt is made to use the information at other locations.'Estimates of condensation are apt to be very local indeed, indicating the need for more and better observation of leaf wetness.

Knowing the incoming radiation ($K\downarrow + L\downarrow$) is only half the story for determining the surface energy balance. Short-wave radiation reflection, $\alpha K\downarrow$, depends on local surface reflectivity, albedo (α). Longwave upward radiation, $L\uparrow$, depends on surface temperature, which may differ much from screen temperature if the soil has a poor thermal conductivity k (Beljaars and Holtslag, 1991). So upward radiation depends on the local terrain parameters α and k. The complete radiation balance depends also on slope and height above sea level, and is therefore much more local than ($K\downarrow + L\downarrow$). Because of advective mixing, air temperature data, when radiation-screened and averaged over a few hours, are rather consistent across a certain distance.

Maximum temperatures, which occur around midday when convection is strongest, allow for some degree of trust in regional forecasts, especially in the summer.

However, attempts to transfer temperatures between locations in a manner similar to that used for wind speeds (McNaughton and Jarvis, 1984) have not been very successful because temperatures continue to be partially linked to small-scale characteristics of their primary source, the radiation-exchanging ground. Therefore, spatial temperature extrapolations have been empirically corrected using estimated net radiation (Gillespie and Barr, 1983). On the other hand, minimum temperatures vary greatly from location to location due to a variety of factors, including the position and vegetation of the soil heat flux at night, which makes up a sizeable portion of energy balances. The katabatic drainage creates additional difficulties in mountainous terrain. Due to the practical necessity of knowing the minimum temperature and the fact that no station can accurately reflect a complicated terrain, modeling attempts strive to combine small-scale terrain data with large-scale advective flow modeling (e.g., Avissar and Mahrer, 1988).

Surface cover, soil type, and soil temperature all affect each other. Therefore, only their changes, such as the observed diurnal course, may be reasonably inferred from observations made at remote stations, but that information is still very helpful. Because they are so reliant on the local surface, vegetation, and radiation conditions, station observations of soil moisture, leaf wetness, grass-minimum temperatures, and pan evaporation cannot be extrapolated with any real reliability. Agriculture depends heavily on soil moisture, although the value of the last two factors is under question. We must monitor soil moisture as a remainder term in the water balance equation, but until small-scale availability and reliability of satellite information significantly increase, and we demand that the data for such monitoring processes should be dependably available.

The gradients in meteorological parameters that emerge where there are geographic complications—sea coasts, big lakes, mountains, and urban agglomerations—must be taken into account when creating representative statistics. Changes in turbidity and cloud cover must be taken into account for radiation. Local wind direction observations may need to be used carefully for the wind depending on local circulations. Anyhow, the significant gradients should be monitored when choosing the station locations. If a region is only capable of supporting, say, three excellent stations, they shouldn't all be situated at the shore or at the base of a mountain range. The use of remotely sensed data is another method for getting a dense spatial distribution of agrometeorological parameter data. The capability of obtaining spectral data from several currently active satellites with ground (pixel) resolutions ranging from around 20 m (SPOT, LANDSAT, ERS) to roughly 1 km is available (NOAA). Due to its frequent observations (twice daily for each place) and the 'convenient' cost of its data, the last satellite, in particular, seems to be extremely attractive for this purpose.

The synopticity of remotely sensed satellite data, which enables the simultaneous collection of environmental data at several spatial scales, is a significant benefit. This characteristic can be used to cover the surrounding land surface with climatic or weather measurements, which are typically gathered locally at ground stations. This is crucial for many operational applications, and recent research has shown that surface temperature estimation can be done with NOAA-AVHRR data. For instance, a hybrid data set made up of AVHRR-NDVI and thermal data has been used to extend with high-precision climatic estimates of mean temperatures and the lengths of arid and cold seasons over the region of Toscana (Italy) (Maselli et al., 1997). Another study found that using kriging, AVHRR thermal data may be used to interpolate daily air temperatures (Petkov et al., 1996). Utilizing the density of estimation points the AVHRR sensor offers (one per 1.1 km in both directions), which could never be economically obtained by conventional surface-based observation, was used to supplement ground information in both cases. The inability to interpolate a number of characteristics over typical synoptic network distances, 50 to 100 km, seems to justify the frequently expressed desire for mesoscale agrometeorological station networks. The agricultural community will require local stations to measure temperature, humidity, leaf wetness, two soil temperatures (say at 0.05 and 0.2 m), and precipitation. Global radiation may also be required, if there is increasing demand for it and a decrease in price. The primary requirements for weather information in agriculture can be met if synoptic stations can increase the utility of their agrometeorological information by measuring wind with adequate precision (gustiness included), providing global and diffuse radiation and additional soil heat information, and so on.

TOPO-CLIMATOLOGICAL EFFECTS ARISING FROM LANDSCAPE VARIATIONS

The typical scale used to depict climate is the fundamental network of forecasting stations, averaged over lengths of at least 50 kilometers. We will mainly focus on landscape variations at meso- and topo-scale, say at least a few kilometers wide, to interpolate over these distances. Following is a helpful breakdown of landscape influences:

- effects of variations in the surface's and soil's physical characteristics (slope, roughness, and moisture content), for instance, on the energy flux partition;
- Local circulations resulting from the effects of (a), and their thermal effects;
- influence of hills, mountains, and other natural or man-made barriers on regional and local air circulations.

The categories overlap, for instance, a long, deep forest serves as both a physical wind barrier and a source of a unique climate close to the ground. A densely populated neighborhood with tall structures will simultaneously act as a physical barrier and change the surface energy balance in the vicinity and create local circulations by storing heat in its dry rocky surface and trapping sunlight in its streets.

Effects of Slopes on Incoming Solar Radiation

Up until this point, we had usually believed that the surface was essentially horizontal. Daytime climate effects on sloped terrain are mostly brought on by variations in irradiation due to variances in the direction to the sun. Effects at night are more dependent on cold air fluxes, and in this case, elevation differences are more significant. Dependence of the Sun's position (latitude, season, and time of day) and slope orientation on direct-beam illumination of a slope (angle, azimuth). The findings can be summed up in a few straightforward sentences:

- In order to maximize irradiation, slopes with an equator-directed component must be steeper the further they are from the equator; this effect is more pronounced in winter than in summer;
- The time between apparent local dawn and apparent local sunset (also known as the "apparent day length") on slopes with a poleward component reduces with increasing slope and may even be nil during some of the winter;
- The morning (for easterly slopes) and afternoon are the peak input times for slopes that are not meridional (i.e., pure N or pure S) (for westerly slopes).

- Tropical slopes receive radiation for a portion of the year as though they were equatorward and for the remaining portion as though they were poleward.

It is crucial to realize that the overall amount of radiation from the Sun when it is hidden by clouds may be comparable to that of direct sunlight. Many solar energy projects would be ineffective if it were not the case. The result of clouds (and turbidity) is an increase in diffuse radiation and a more uniform distribution of that diffuse component over the sky canopy by blocking out some of the direct sunlight. As a result, orientation and slope lose significance as cloud cover increases. Geiger cites data that show the north face of a mountain at S0°N with a 30° slope gets, on average, the following percentage of the radiation that strikes the south face:

- Direct radiation only ---- 2 percent(December) ---73 percent (June)
- Direct and diffuse---- 32 percent (December) ---94 percent (June)

Because small aerosol particles undergo Rayleigh-type scattering when the sky is bright, radiation is primarily dispersed forward in such conditions. Then diffuse radiation is strongest close to the Sun and is least light while facing 90 degrees away from the Sun. Because larger, Mie-type scattering droplets are present in clouds, the isotropy of radiation is increased. For higher levels of cloudiness, the ground slope is therefore less significant for the overall intake.

If topographically generated clouds have semi-permanent placements, the radiation environment in mountainous regions can change locally. With latitude, the impact of slopes on vegetation changes dramatically.

In the tropics, where radiation is typically adequate and humidity is occasionally insufficient, vegetation may be lusher on slopes that face poleward. It is crucial for agriculture to pay attention to slopes since it is more critical to take advantage of the direct sunlight that occurs on equator-directed slopes in temperate and arctic climates, as well as more cloudy locations. In Germany, only slopes facing the southwest are suitable for growing grapes. This is especially true on steep slopes where, in the fall, the peak of the day's heat coincides with the maximum irradiation.

Soil Temperatures on Slopes

Increased soil temperature of the bare soil, or scant vegetative cover, during germination and early growth of annual crops, is likely the most significant agricultural effect of the link between slope and direct sun input. Because turbulent buoyancy aids the air's ability to expel heat on sunny days, exposure and slope inclination have a greater impact on the daily maximum soil temperature than on the daily maximum air temperature. The total leaf area (LAI) should be at its highest when solar radiation is at its peak since photosynthesis in plants depends on the collection of solar radiation by leaves. Many crops are unable to completely utilize the high radiant input of midsummer due to a climatological lag of temperature (which affects leaf development) in maritime environments, especially in the higher latitudes. As a result, early germination and growth are particularly advantageous, and this can be accomplished by taking advantage of the warmer soil temperatures found on slopes that are slanted up toward the Sun. The results of a groundbreaking study by Wollny are shown in the table below, which highlights the significance of slope-effects on soil temperatures (1878). The time of day affects the temperature variations between slopes of various aspects and degrees.

Table 21: *Soil temperatures at 15 cm depth in 15° slopes facing eight main directions.*

Direction of the slope								
Month	**N**	**NE**	**E**	**SE**	**S**	**SW**	**W**	**NW**
May	10.7	10.9	11.2	11.4	11.3	11.3	11.0	10.9
June	20.4	20.6	20.8	21.4	21.4	21.4	20.9	20.6
July	18.7	18.9	19.1	19.4	19.3	19.2	18.9	18.7
August	19.2	19.4	19.8	20.4	20.5	20.5	19.9	19.2
September	12.3	12.5	13.2	13.7	13.9	13.7	13.2	12.7
Temperatures in °C								

Source: Wollny (1878).

The observations in the table were made on 12 clear winter days in the Gobi Desert (42°N) on a 23-meter-high sand dune; the angle of slope is not indicated, although for a stable dune, it would be significantly less than the'maximum angle of repose of a sand dune (54°). The soil temperature at a depth of 2 mm reached 0°C on the summit and 22°C on the south slope, while the air temperature (measured in a shelter) remained below freezing around noon. The average noon temperatures at 40 mm depth were 2°C and 10°C, respectively. Geiger provides diagrams of complete diurnal courses, which are just as significant for agricultural uses as mean daily values.

Achieving greater temperatures for brief periods of time during the day may be crucial at some stages of growth.

Effects of Slopes and Hills on Airflow

Mountain influence on tropospheric airflow has received extensive research as a micrometeorological issue (e.g., WMO-TN No. 34). Mesoscale airflow is the subject of our interest, and two key issues warrant explanation. The first is how certain aspects of hilly terrain alter larger-scale airflow. The second is the creation of regional circulations, especially as a result of solar radiation's impacts on slopes. The height of the hills themselves, the height differences between ridges and valleys, the shapes (such as the ratio of a valley's width to the height of the surrounding slopes), the convexity or concavity of mountain ranges, the extension of high ground (or "spurs") down to the surrounding plains, and the orientation of all these features relative to one another are important considerations when examining the effects of mountainous terrain on wind-flow. The slope radiation effects outlined above, for instance, are not significant in studies of gale damage to mountain forestry because buoyancy effects become secondary in strong gusts.

Results of wind tunnel and model studies, such as the windbreak barrier flows, can teach us a lot. An oncoming air current will flow smoothly from the slope's base all the way to the crest (of 'infinite' cross-wind extent). Particularly when hills rise abruptly from a plain, an upwind "bolster" eddy will be present at the base of a steep ridge. According to wind-tunnel studies (Frost, 1973), the oncoming flow separates at about 1.0 H upwind in the case of a block with a rectangular cross-section or a vertical step (a shape like a vertical cliff face presents to the wind). The bolster eddy then extends to about 0.6 H up the face, where the main flow once more impinges on the face. This upwind bolster eddy appears to occur at a crucial slope angle of around 25°. We can assume that in most circumstances, oncoming wind can be expected to go smoothly up these slopes because the majority of arable farming is done on slopes of less than 10° and the majority of pasture farming is done on slopes of less than 20°. For forestry, where gale damage is a significant economic factor, and terrace cultivation, the situation is different. At sharp edges, air traveling over a ridge may divide (i.e., cease to follow ground contours). The upwind edge of the ridge top area is the first possible separation point; the area directly downwind of that edge may experience wind speeds double the average and intense turbulence (Bowen and Lindley, 1977). If the slope of the ground is sufficiently downward, the wind flow will separate beyond the ridge. Thermal convection has an impact on this; according to Scorer (1978), convection increases the likelihood of separation over leeward slopes but inhibits it over windward slopes. A small upslope wind may be created by sunlight on a lee slope, which can help to separate the main flow and create a lee eddy.

Vegetation on the lee slope is typically protected by separation. Geiger cites Kaiser's findings that at very shallow lee slopes (2° to 4°), there was no propensity to flow detachments at speeds up to 4 m s-1, but at 5 to 10 m s-1, the lee was clearly protected. Large-scale winds' interactions with valleys result in contrasting flow patterns. Because of canalization, the wind speed is greater than the regional wind when the direction of the large-scale wind is similar to the line of the valley. Unexpectedly, the larger valleys—rather than the tight valleys of heavily serrated terrain—seem to see the most noticeable gains in speed. The larger significance of frictional drag in narrower valleys can be used to explain this. If the distance between successive ridges and the depth of the valley floor is just right, wind may also infiltrate down the lee slope and across the floor when the local wind blows over the valley's line. Other possibilities include the formation of roller eddies. Hutte (1968) describes the variations in the approach angle of a current impinging on an (essentially) infinite ridge as causing changes in the wind pattern:

- The lower slopes experience the most wind speed and turbulence when the wind direction is parallel to the ridge line.
- The mid-slope region experiences turbulent winds with frequent direction changes when the wind direction is 20° to 50° to the ridge line.
- The wind picks up speed from the base of the ridge to the top when the wind direction is perpendicular to the ridge line.

According to experience in the United Kingdom, in truly severe gales the pattern of comparatively exposed and relatively sheltered zones, evident in moderate or strong winds, breaks down and the severe winds surround the entire area, creating damage that is impossible to identify with any topography.

Intense turbulence can be created halfway up a hill's slope in heavy winds. Additionally, when a series of roughly parallel ridges is present, the wind current can be redirected upward by the leading ridge, drop, and reattach to the windward base of the second ridge, causing the most severe damage to forests to occur on the mid-upper slopes of the second ridge.

Local Mesoscale Circulations

There are two phenomena at play in mountain valleys. First, the valley and its differential heating and cooling produce flows. Second, as was already mentioned, the valley walls mechanically restrict air circulation in general. The circulations described below may be a bit obscured by powerful large-scale winds. Due to buoyancy factors, upslope breezes will typically form on valley slopes that face the Sun. The valley's direction is obviously important. In a north-south valley, both east- and west-facing slopes receive an equal amount of sunlight during high sun hours. Then, on both slopes, upslope winds will occur when large-scale wind is weak. There are three ways that these air currents can start to flow backward. First, the center of the valley can have a balancing descending current. Second, on non-sunlit slopes, buoyant up-flow can be countered by downslope breezes. Finally, as valley bottoms often rise from the plains to the center high portion of the mountainous area, there is typically an along-valley flow and three-dimensional return flow is possible. The vertical extent of the valley circulation phenomena is typically 1.5 to 2 kilometers. At night, all slopes, especially those that face the sky directly, radiate heat into the atmosphere, cooling it. This cold surface air then behaves like water, flowing downslope.

Barriers of all sizes, including valley narrowing, forests or tree-lines, and particularly local peaks upwards of which cold air "lakes" form, can alter this flow. The katabatic wind's downflow typically moves at a speed of little more than 2 m s-1.

The air rushes down the slope until the cold air "lake" is depleted, although in rugged terrain, "air avalanches" can happen when cold air dams are avoided or overtopped. The avalanche will happen again when the store is restocked, and these bursts of icy air frequently happen at fairly regular intervals. Since this form of severe katabatic flow is generated by temperature differences, Scaetta (1935) saw one of these flows nearby Lake Kivu in Central Africa, demonstrating that such phenomena are not limited to high latitudes. For nocturnal cold air drainage, very mild slopes are adequate: Geiger reports a frost issue brought on by a slope of 1:200, or a 0.3° slope angle. Any onlooker would perceive this terrain as level. At the topo-scale, every terrain features so-called "frost hollows," which are small depressions that can be either naturally occurring or purposefully produced and have limited or no openings for cold air to enter. In relation to lesser scales, this is examined.

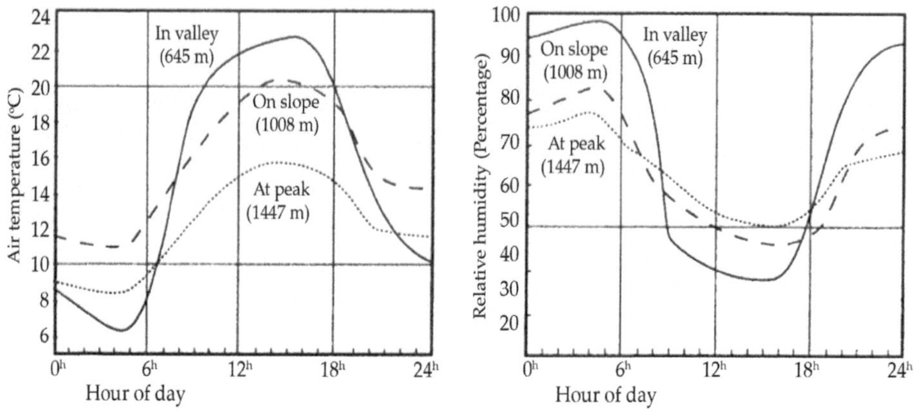

***Figure 43**: Diurnal air temperature and humidity variations at screen height*

Source: Geiger, Oke (1987).

The graphic above, which provides information on diurnal courses in a valley and further upslope, is a valuable summary of the mesoscale effects of valley flows for agricultural planning reasons. At night, a lake of cold air forms at the valley's bottom, but cold air still exists on the top, and if the peak is high enough, it will have a lower temperature than the valley. The thermal belt is a warmer intermediate zone that is normally located between 50 and 300 meters above the valley bottom, depending on the local geology. For obvious reasons, this thermal belt is a preferred zone for agricultural and other uses, such as habitation, in temperate and polar climates. The sea breeze phenomenon is another well-known type of mesoscale circulation. Since E and G are large at sea or over a lake that is quite deep, the vertical heat flux is minimal. In contrast to the advected air, the water's surface temperature is therefore constant. The pressure at 1 km height is, therefore, higher on the warm-surface side of a shoreline than on the cold-surface side (i.e., during the day over land and at sea during the night). This creates a circulation that causes daytime sea breezes and nighttime land breezes to form near the surface.

According to Oke (1987), this phenomenon has a diurnal course and is typically limited to areas that are between 20 and 50 kilometers inland from the shore. There may be unusual air movement as a result of these coastal circulations. Van Arsdel (1965) noted pine blister rust 15 km inland from a lake, perhaps brought on by spores released at dusk near the coast, but no infections were present in the intermediate region. Evidently, the spores that were released at the coast were caught by the land breeze and transported by the return circulation at a height of 1 km to the downward flow at the inland circulation limit! Anywhere a mesoscale surface difference exists, further significant mesoscale circulations are feasible. Cities and desert oases both have mesoscale circulations.

CHAPTER

7 Agrometeorological Management

CONTENTS

OVERVIEW

Agriculture can benefit from a meteorological understanding of energy exchanges and transportation at the surface. The techniques discussed in this chapter are intended to control the environment close to the ground, in part by modifying the moisture content and soil temperature in the top soil layers, such as by mulching, and in part by taking action above ground, such as by providing shade or using windbreaks. Due to the themes' tight interrelations, it is somewhat arbitrary to divide them into separate topics. Land plants exist in two different environments: the topsoil and the air near the ground. For the majority of plants, before the seeds have germinated, the soil is the only medium available. For root elongation, tuber development, and seed germination, soil temperature can be a limiting or crucial element. The amount of carbon dioxide that passes from the soil into the layer of plant-air and into the atmosphere is also impacted by the decomposition of organic matter inside the soil.

PRINCIPLES AND APPLICATIONS OF AGRICULTURAL METEOROLOGY

Agrometeorology has both a strategic and a tactical role. The strategic function is involved in the evaluation of long-term resource utilization in the growth of crop diversity. The tactical role is more focused on making immediate, field-level decisions that have a direct impact on crop growth and development. Agrometeorological data can assist farmers in practicing sustainable, high-quality, more lucrative agriculture with fewer risks, lower costs, and less environmental pollution and harm if it is communicated to the correct client and utilized (Rijks and Baradas, 2000). A significant portion of the annual crop losses in global agriculture are directly related to meteorological events like frost, hail, storms, flash floods, and drought. The weather has a significant impact on harvest and storage losses as well as those brought on by parasites, insects, and plant diseases (Mavi, 1994). When weather information that is specifically tailored to agricultural needs is easily accessible, it greatly aids in making short-term adjustments in daily agricultural operations that minimize losses brought on by unfavorable weather conditions and enhance the yield and quality of agricultural products. Additionally, specific meteorological information offers recommendations for long-term or seasonal planning and the choice of crops best suited to expected climatic circumstances (Newman, 1974; Ogallo, Boulahya, and Keane, 2000).

The majority of decisions made in cattle companies have a long lag between them and their impact. A few choices have a three- to four-year impact on the product (Plant, 2000). A very effective strategy for managing climate risk and boosting livestock productivity is a long-range projection. Governmental economic policy can be significantly influenced by seasonal climatic projections. For instance, if a severe drought was predicted, economic growth would be lower than anticipated. Monetary policy could be loosened in order to sustain growth targets by seriously considering the prediction. Through the development of methods based on a solid interpretation of meteorological knowledge, agrometeorology also finds use in other fields. The effective use of herbicides, insecticides, and fertilizers, as well as the shelter from wind or cold, shade from excessive heat, anti-frost measures, including site selection, anti-erosion measures, soil cover and mulching, plant cover using glass or plastic materials, artificial climates of growth chambers or heated structures, and animal housing and management, storage and transport climate control, are just a few of these. Agrometeorological models can be used for successful land use planning, identifying the right crops for a location, risk analysis of climatic hazards, farming profit estimates, production or harvest projections, adoption of farming practices, and the selection of efficient farm equipment.

Solar Radiation

Plant growth requires solar radiation. Sunlight is absorbed by plant leaves, which use it as a source of energy for photosynthesis. The amount of leaf surface area, also known as the leaf area index, determines how well a crop can absorb sunlight. A crop's capacity to absorb sunlight is at its peak when the canopy is fully developed. Crops' ability to absorb sunlight can be hampered by agronomic variables including weed competition, insect feeding, or leaf diseases, which can diminish leaf surface area. Theoretically, crop productivity will rise together with the amount of radiation energy gathered. The temperature of the leaf surface rises when plant leaves utilize solar energy for photosynthesis. To cool the surface of the leaf, plants react by releasing water through their stomata. Carbon dioxide from the atmosphere is absorbed by plant leaves, and water is absorbed by roots. The energy plants require to transform carbon dioxide and water into carbohydrates and oxygen is provided by sunlight. In order to enhance crop biomass and support vegetative and reproductive growth, photosynthesis produces carbohydrates. Photosynthesis can only take place in the daylight since it requires sun energy. The amount of water vapor in the atmosphere has an impact on the amount of solar energy that reaches a crop. By reflecting solar radiation back into space and keeping it from reaching crops, clouds limit solar radiation. Future cloud cover will gradually decrease as the amount of greenhouse gases in the atmosphere rises, which will contribute to greater global warming. By sowing their crops as early as is practical each spring, farmers can enhance the likelihood that their crops will be able to catch solar light. For instance, compared to wheat sown on May 30, whose days lengthen for only three weeks before shortening once more, wheat sown on May 3 will experience seven weeks of lengthening days. Generally speaking, earlier sowing can benefit crop output.

Soil Moisture and Crop Water Use

Numerous factors, including crop growth phases, crop rooting depths, soil water availability, precipitation levels during the growing season, and climatic elements such as solar radiation, humidity, temperature, and wind, influence how much water a crop absorbs and uses. The amount of water a crop uses for cooling and growth is known as crop water use. Crop water use can be calculated daily, weekly, or seasonally. Evapotranspiration (ET) is the name for how plants utilize water. ET is the result of water evaporating from plant and soil surfaces as well as water required by plants for transpiration and growth. Water lost to the atmosphere through a plant's stomata, or pores, which are typically found on the undersides of leaves, is referred to as transpiration. To reduce heat stress, plants produce water vapor through their stomata, which cools the surface of their leaves. Evaporation is typically only considerable following precipitation when the crop canopy and soil surface are both wet. Water evaporation from soil often occurs after the top two to six centimeters (cm) of surface soil have dried. Additionally, as the crop canopy closes to entirely shade the soil surface, evaporation from the soil's surface is reduced. Nearly all of the ET at full crop canopy comes from crop transpiration. When soil water is not a limiting issue, the ET rate is at its maximum. The temperature of the leaf is kept lower than the ambient air temperature through stomatal water transpiration, followed by evaporation from the leaf surface. The guard cells found in stomata control transpiration water loss by opening and closing the stomatal opening. To keep leaf surfaces cool, transpiration is most effective when the stomata are fully open. To lessen transpiration when plants are under water stress, the stomata partially close. Transpiration is influenced by humidity, which is a measurement of the amount of water vapor in the atmosphere. When air humidity is low, a crop transpires more quickly. In a crop canopy, there is frequently a layer of still air close to the leaf surface. The cooling effect of this layer of still air extends from the leaf to the atmosphere. The still air layer's thickness affects how much water plants lose through transpiration. The layer will be disturbed by the wind; as the wind picks up speed, the still air layer thins out, increasing the pace at which leaves are losing moisture to the moving air. Crops draw water from the earth using their root systems. The amount of water in the soil, the stage of plant growth, and the effective rooting depth all have an impact on how quickly and how much water a crop absorbs. A specific level of moisture is required for annual crops to start germination and progress through the vegetative and reproductive growth phases to the point where seed can be generated. A crop needs at least 100 millimeters (mm), or four inches, and frequently closer to 125 mm, or five inches, of water to develop from germination to the reproductive growth stage for wheat, barley, and canola. The amount of moisture required varies depending on the climate, as crops require less moisture for transpiration on cool days than they do on hot ones.

At the tillering stage, cereal crops require roughly two to three millimeters of water per day; at the stem elongation stage, they require roughly three to five millimeters per day. The amount of moisture required is roughly five mm each day when the temperature is above 25 C. A grain crop will consume between 20 and 35 mm of water in one week on warm days at the stem elongation growth stage, depending on the climate. Under ideal circumstances, water use for cereal crops at the heading stage is seven to eight millimeters per day. This indicates that mid-June to late July or early August is when water use is at its highest. If crops don't have enough moisture, they won't be able to use as much water, which will result in a considerable drop in output.

Water demand for crops stays high as they transition from vegetative to reproductive growth. After heading, cereal crops and canola in the flowering development stage will continue to consume seven to eight mm of water per day until seed filling under ideal growth conditions. Crop water use decreases as grain filling nears completion and drops very quickly as plants mature. After the vegetative development stage, wheat can generate five to eight bushels per acre (bu/ac) for every 25 mm (one inch) of water consumed, according to research conducted in Alberta. For every 25 mm of water utilized following the vegetative development stage, barley may generate seven to ten bu/ac while canola can produce 3.5 to five bu/ac. A deficit develops when plants are unable to absorb enough water from the soil to meet their needs. C3 crops block their stomata to prevent dehydration, which lowers photosynthetic activity. Crop yield potential likewise declines as the water deficit and photosynthesis both rise. The majority of temperate crops, including wheat, beans, potatoes, and trees, are C3 plants, which are also the most prevalent and have the highest photosynthetic efficiency. When there is a severe shortage of precipitation and soil moisture, a drought develops, which significantly lowers crop productivity. When crop yields are lowered by at least 50% below the long-term average yield, a drought is typically deemed severe. A severe drought typically results from both higher-than-normal temperatures and less precipitation than is typical.

Temperature

Cardinal temperatures are the lowest, ideal, and maximum temperatures at which all crops' growth processes are impacted. For instance, several plant functions like germination, vegetative growth, root growth, water intake, photosynthesis, respiration, flowering, and seed development require minimum temperatures. Processes in the plant will halt at temperatures below the minimum. Processes in plants operate most efficiently at optimal temperatures, while they halt at maximum temperatures. In the southern Prairies, heat stress and water stress are frequent, although they also happen occasionally in all agricultural regions. They could happen one at a time or all at once. Due to insufficient soil moisture, plants that are under water stress must close their stomata in order to meet their needs for transpiration. This raises the temperature of the plant. The maximum temperature that might cause heat stress depends on the type of crop and its development stage. The normal range of ideal temperatures for photosynthesis in wheat is between 15 and 30 ^{0}C. Between 30 and 40 ^{0}C, heat stress is more pronounced; over 40 ^{0}C, the photosynthetic processes can irreparably fail. Different growth phases might be impacted by high temperatures. For instance, high temperatures five to ten days prior to anthesis, when pollen is generated and viability might be severely impacted, have an effect on wheat production.

Plants that are subjected to low temperatures below 0°C may experience low temperature or chilling stress. Plants experience freezing stress when they are exposed to temperatures below 0 C, which can have a negative impact on crop growth and output. The ability to withstand lower temperatures is greater in plants that have gone through periods of cold weather prior to a frost. For instance, seedling wheat and canola that have been acclimated to low temperatures may frequently withstand temperatures as low as - 6 0C. The better we can plan and create more sustainable farming techniques, the more we learn about the interactions between our crops and our unpredictable weather and environment. All agronomy disciplines must be taken into account, including the sorts of crops that can be cultivated and the control of fertilizers, weeds, insects, and diseases, all of which are influenced by changing weather conditions. Then crop production techniques can be gradually modified to reduce the consequences of harsh weather. Crop production will be successful throughout a larger range of changeable temperatures and water availability in the future thanks to advancements in crop genetics, improved agronomic techniques, and better long-term weather forecasts.

MANAGING AGROMETEOROLOGICAL FACTORS

Soil Cultivation and Treatment

The use of activities, techniques, and treatments to safeguard soil and improve its functionality is known as soil management (such as soil fertility or soil mechanics). It entails soil amendment, soil conservation, and ideal soil health. To keep agricultural land from degrading in productivity over many years, both nonorganic and organic kinds of soil management are required in agriculture. Growing soil is a never-ending chore. Because both weather and agricultural practices can cause the soil to deteriorate again, it is usually only a transient improvement in soil climate. Eventually, it will need to be repeated. Only subsoiling might have a more lasting impact, especially if the subsurface is impermeable. Historically, the main purposes of soil cultivation were weed control and the establishment of a seed bed. Even though burning or applying weed killers can take the place of mechanical weed eradication, seed bed preparation is still crucial because there will be much less time after sowing to improve the soil's condition. Cultivation may make the soil more porous, which will afterwards improve its aeration and temperature as well as its drainage. The climatological conditions in soil and air, evapotranspiration, and the air temperature near the ground and its daily change are all impacted by soil cultivation. The focus of this discussion will be on soil condition changes caused by variations in surface albedo, by changes brought on by a thin "mulch" soil layer, and as a result of surface reshaping as ridges and furrows.

Surface Color on Soil Temperature

Differently colored soil surfaces reflect shortwave solar radiation with varying coefficients, and as a result, different amounts of solar energy are absorbed by the surface. Therefore, changing the surface hue will vary the soil's net radiation. We have limited control over long-wave interaction because infrared albedos are always close to 1. In agriculture, colorful surface coatings are used to warm the top soil layers or to prevent high temperatures that could harm young plants and seedlings. When the entire spectrum is taken into account, the extreme conditions—such as black cotton soil, soil covered in soot or carbon black, and, in particular, fresh snow—are the only ones that exhibit noticeable changes in the reflection of diffuse radiation from natural surfaces. Temperature changes in the soil can be impacted by surface treatments. Dark materials can be added to the soil in temperate climates with a cold season to raise the soil temperature in the spring for earlier germination. There are methods to change the soil's albedo;

Mulching

Mulching inhibits the vertical passage of heat and water vapor by applying or creating a layer of soil cover. Mulches differ in their porosity to vapor, heat conductivity and capacity, and surface albedo, all of which are significant. Topsoil, cut or harvested vegetable material, repurposed surface debris, and manufactured materials are all acceptable ingredients in mulches. In various temperature conditions, a thick layer of snow acts as a mulch to protect the subsoil from extreme cold. Since it applies itself in the early winter and removes itself in the spring, it also has the practical benefit over other mulches. The non-convecting air layer beneath a heat-conducting cover, such as aluminum foil, works as a heat flow barrier since mulch has low thermal conductivity. As a result, the subsoil is protected from short-term temperature extremes, such as midday or summer heat and nighttime or wintry cold. This means that daily and yearly changes in surface temperature are poorly conveyed downwards. Mulching reduces soil evaporation since there is less vaporization energy available. Incoming radiation heats up the surface if it can't penetrate the soil effectively. Similar to this, the surface cools down more at night if the emitted surface radiation cannot be replaced from below. Because of this, the diurnal temperature variation is higher immediately above a mulch than it would be on the bare soil's surface. This variation is also influenced by the mulch's albedo. One drawback of mulches is that young plants suffer from unusually high daily temperature changes as soon as their tips grow above the mulch. The mulch may need to be removed at that point in the growth process. The selection of a mulch, as well as its conductivity and albedo, undoubtedly deserves careful consideration. Of course, employing mulches has potential drawbacks. For instance, it can be quite detrimental when soil microbe activity is restricted due to a decrease in temperature or a reduction in nitrogen. In the latter instance, the organic mulch has a higher concentration of bacteria than the soil itself. Mulches can also change the microfauna in terms of composition.

The increased danger of crop frost as a result of mulching is a significant adverse effect. In actuality, removing loosely packed vegetative litter from a plantation or orchard is the first step in lowering the risk of frost because such a layer transmits heat very poorly. In order to improve the conductivity of soil heat flow, another frost-prevention strategy involves compacting the soil to increase the contact between soil particles. Perhaps mulching should be viewed as an inversion of soil compaction. Mulching has additional negative impacts. Reduced wind erosion is a benefit and a common rationale for mulching, as it provides both surface shade and longer-lasting soil moisture. Additionally, the impact of cover on runoff and soil percolation may be significant.

Surface Geometry Effects on Temperatures

These include results of minor topographical features, such as mounds, ridges, furrows, and trenches, having vertical dimensions of no more than a few meters. Their primary impact is to change the amount of solar radiation (especially direct radiation) that strikes the surface, which has an impact on the temperature and hydration of the soil.

Thus, the germination and emergence of a wide variety of crops can be modified. Similar to those addressed in the preceding chapter, there are mathematical correlations between direct sunlight and the slope and aspect of minor ridges and furrows during the day. Their effects, in particular, are greatly influenced by their direction in relation to slopes and the position of the sun. Because of the expansive sky view and the (relatively) great distance to deep soil heat, the summits of ridges are colder in the morning than at ground level. After then, depending on orientation and also because reflected radiation may still be absorbed at the other side, sunny slopes heat up significantly (radiation trapping).

Crop Management and Layout

Particularly when there is low PAR at high latitudes, direct accessibility of Photosynthetically Active Radiation (PAR) to leaves becomes crucial. Depending on the type of crop and its stage of growth, excessive heat or radiation may also pose a risk in other situations, such as when it causes an excessive amount of water stress. Such conditions need the use of artificial or vegetative cover, which also has the advantage of reducing wind erosion.

Spacing of Crop Rows

When a crop entirely covers the ground, the crop itself has a significant impact on the surface absolute values of temperature, humidity, and winds as well as their vertical gradients. On horizontal terrain, the spacing between row crops and seasonal variations in the crop-to-row spacing have a significant impact on the amount of sunlight that reaches the real soil surface. Always keep in mind that the ratio of direct to diffuse sunlight affects radiation phenomena when talking about this and similar issues. Some orientation-related differences disappear in cloudy conditions. The production of individual plants and planting density both have an impact on crop output per unit area. If densities are higher than ideal, the light's intensity in the photosynthetically active region of the radiation spectrum is too severely reduced even though it can still reach lower portions of the crop. Additionally, when crop density increases, air flow within the crop diminishes, which may lower the CO2 concentration near leaves and restrict photosynthesis on bright days with weak winds.

Shading, Cover Crops and Weeding

Both agriculture and animal husbandry use the technique of shade (i.e., interposing some sort of screening against sun radiation). Providing a vegetative cover can also have large-scale effects by preventing wind and water erosion. It is somewhat difficult how shade affects the microclimate. The surface of trees that provide shade or roofing reflects a sizeable portion of the total global radiation. The portion of the sun's radiation that is not reflected is partially absorbed by trees that provide shade or by roofs, and the remainder is transmitted. The majority of the energy absorbed will be converted into thermal energy in the presence of a roof composed of dead material that shades the area. Remember that green leaves produce shading, which modifies the spectral composition of solar radiation while reducing the total amount of solar energy reaching the surface of the shaded plants. The primary alteration in the radiation' is brought about by the preferential absorption of light with a wavelength shorter than 0.75 m, specifically in the visual range's blue and red parts.

Additionally, shade will tend to lessen the horizontal and vertical movement of wind. The daily mean temperature of the air and soil may be a little lower depending on the region and the kind of shading, but the diurnal variation of the daily temperature is always less. Because shade trees drain soil water and evaporate rainwater that has been caught in their leaves, the overall effect of shading on soil moisture is complicated. When it rains frequently and in large amounts, the effect of shading on moisture retention is unlikely to be significant; nevertheless, when it rains infrequently, living shade can frequently be damaging. As we progress from simple devices, whose primary goal is to reduce the intensity of short-wave radiation reaching the foliage, to complex devices with more purposes than simply shading - ranging from removable low glass or plastic structures (cloches) to the controlled environment of fully equipped glasshouses - the complexity of the changes in microclimate as a result of the use of shade or cover become more pronounced.

Wind Shelter

Wind Shelter and Microclimate

When there is "zero" wind, as is the case in glasshouses or areas with little glass cover, vegetation grows lushly. Typically, as the wind speed increases to minor values, such as 1 or 2 m s-1, vegetative growth slows down. According to the net heat exchange by convection between the plant surface and the air, irradiated plants' temperatures vary. Condensation causes it to rise while evaporation causes it to fall. On a somewhat larger scale, if transpired water vapor is to be evacuated and an appropriate supply of CO2 is to be maintained for photosynthesis, an adequate degree of air motion ('ventilation') is necessary within a crop stand and surrounding plants. Additionally, the sluggish, wet air in humid areas typically creates ideal circumstances for disease and insect growth and plant attack. The best wind speeds depend on the vegetation type and the overall climate regime. Since plants can only pass water at a certain rate, the ventilation must be adequate, not more. Stomata close due to internal water scarcity brought on by excessive transpiration when wind speeds are too high. Because it is typically cloudier in temperate maritime regions, the level of air temperature in the sheltered zone (up to 10 H or 15 H) during the daytime is typically 0.5-1°C higher in warm temperate regions and 2 to 3°C higher in those regions. Due to the formation of a body of still air that is partially cut off from the significantly warmer air that can be found farther above the surface on clear "radiation nights," temperatures behind a belt may be lower than in exposed terrain during these times. On other nights, there are no discernible differences between the minimum temperatures in covered and unsheltered areas. The comparatively high daytime soil temperatures in shelters have an interesting dynamic effect in that they suggest that greater soil heat will be accessible at night to prevent ground frost. However, due to the effects of radiation exchange with the belt rather than the sky, the risk of ground frost in shelters is higher, with the exception of very close the belt (3 H distance). Near shelters, relative humidity and vapor pressures are higher despite higher temperatures, however, the latter's increase is rarely greater than 10%. Due to higher humidity and less wind, there is also a disproportionately high amount of dew found close to cover. The main reason for the difference in soil temperature between shaded and unsheltered plots is less evaporation. However, due to the fact that damp soil has greater heat capacity than dry soil, changes in soil temperature brought on by wind reduction are uneven and limited, depending on the kind of soil and ground cover. Additional issues can occur next to a shelterbelt because of shading, which in turn depends on the type and direction of the belt. The net difference for belts that are North-South oriented is little because higher afternoon reflection, or vice versa, compensates for morning shade.

Windbreaks against Damage and Erosion

Other than the damage brought on by gales ($U > 14\ m\ s^{-1}$) and hurricanes, less severe winds can also inflict harm. Effects become more severe as wind speed at the typical height rises to exceed 5 m s-1 and higher. Reduced mechanical damage and minimizing distorted growth are the sources of many benefits of wind protection. Mechanical damage can result from frictional rubbing between various plants or different portions of the same plant at wind speeds of roughly 5 m s-1. Even at lower wind speeds, especially fragile crops like flowers or horticultural crops are damaged, as sustained gusts over this level can stunt plant growth (including fruit trees). The shape of plants is influenced by higher sustained speeds, which results in the recognizable wind-shaping of trees in exposed situations. Smaller and thicker leaves have a tendency to grow under these circumstances, and the windward side of bushes and trees is naturally pruned. At 10 m s-1, CO2 assimilation only continues at 10% of the rate observed in calm conditions. Plants take on a low, dwarf-like morphology where average winds are more than 12 m s-1. Airborne material (weed seeds, small insects, spores, i.e., agents associated with pests and diseases) tends to drop out of the general airstream and/or be deposited by eddies and concentrated in the protected zone due to the lower forward speed in the zone of influence of the windbreak. Although quieter environments frequently result in better microclimate, better pollination, and more effective application of insecticides and pesticides, it is important to be aware of the drawbacks (increased pest and spore concentrations). Wind has an impact on operations in addition to having an impact on crops. Low wind speeds are better for carrying out overhead irrigation and spraying liquids or dust. Shelters can greatly lessen the harm caused by sea spray and salt particles in coastal areas.

Strong winds blowing over loose soil move particles and, if they don't stick together, produce erosion (no cohesion). The amount of moisture in exposed soil greatly affects wind erosion, which impacts cohesiveness. Low mean annual rainfall areas, especially those with less than 250 or 300 mm, are especially prone to erosion. Shelterbelts can contain wind-driven soil and slow down or stop wind erosion's ability to carry large amounts of material. Shelter is utilized in dry places to lessen the incursion of wind-blown sand and soil. Depending on the type of belt, windbreaks have different effects on sand deposition and sand cover. If the belt is porous, the majority of the sand is deposited in the more still conditions behind the band. Near dense windbreaks, wind-driven sand accumulates on the windward side and peaks at the windward edge of the belt. The belt needs to be densely planted, have a minimum thickness of 30 m, and be made up of partially buried trees like tamarisk and eucalyptus in order to successfully prevent deposition over time. Crop remains left in the field as a mulch can help keep the surface covered and decrease erosion of cultivated arid soil. When the forecast indicates an adverse mix of strong winds and significant evaporation, certain soil-loosening procedures can be postponed in non-arid places to prevent unnecessary wind erosion damage. Shelterbelts can affect soil moisture in temperate climates by affecting snow deposition, which results in buildup in drifts on both the windward and lee side of a belt. When there is enough snow, drifts can also form in the sheltered area; the more open the barrier, the less buildup. Drifts are typically deeper on the windward side. Such snowdrifts can be a helpful boost to the soil's moisture stores in arid areas. However, huge snowdrifts in humid areas typically have drawbacks since they pose a threat to roaming cattle and are difficult to transport. The concentration of meltwater can also greatly postpone the springtime cultivation of the subsurface soil. Snow deposits are frequently more even in protected locations behind a belt, with the associated benefits.

Irrigation and Drainage

The fundamental tenet of irrigation is to supply enough water to the root zone. However, the amount of water needed by a crop varies greatly depending on the crop and where it is in its growth cycle. Nearly all crops need enough water during the flowering season. The crop, local weather, and soil conditions all affect how much they require before and after. The majority of crops can only grow and develop entirely unhindered and produce at their highest levels if high water status is maintained throughout the crop's life. We will only mention a handful of the many risks associated with excessive water use. The first is that water that is added in excess of what is needed to fill the soil to field capacity will either drain away or drink down to levels below the root zone, "leaching" along with it any soluble soil nutrients, including those that have previously been supplied as fertilizer. Another risk is the deterioration of the soil caused by excessive irrigation. When the water table is close to the surface, intensive irrigation has caused "salinity" in sub-arid and desert countries (like Pakistan) (at most a few meters below). A salt crust forms when the more water available dissolves the salt in the soil and some of the saline water rises to the surface due to the high evaporative potential. Additionally, if the uncommon but powerful rainstorms fall on recently irrigated terrain in these areas, both the likelihood and amount of erosion by runoff are exacerbated. The pores of some soils with a high loam content or other fine-structured soil types can become plugged by smaller soil particles in conditions of abundant water in both temperate and tropical locations. It won't be possible to reverse the adverse soil impermeability that results until after harvest. In the interim, it prevents oxygen and carbon dioxide from diffusing through the soil, which may cause the crop's absorbing roots to wither. Additionally, too much irrigation during the period of root development may result in the formation of a root system that is excessively shallow. This renders the crop later on very susceptible to drought. The effects of irrigation are influenced by the actual methods used. Ditching is a traditional low-cost method that makes use of lateral conveyance through the soil. Irrigating the furrows in ridge-and-furrow farming is a useful modification since, on level terrain, this might cause inhomogeneity in water supplies. Flood-irrigation, which is required for rice, is the extreme instance. Above-canopy irrigation, also known as sprinkler-irrigation or spray-irrigation, has relatively high application costs. On soils with rapid drainage, it can be the only method that works. It also momentarily raises the air's humidity as a side effect. Water being sprayed directly on the leaves has another significant negative effect. This has two effects: first, direct evaporation from the leaves cools the environment, and second, induced leaf wetness increases the likelihood of fungus infection. Cooling by 10°C is advantageous because dry leaf temperatures in subtropical climes can exceed 40°C. To avoid dangerously protracted wet-leaf periods, spraying should not go on for too long. It is best to spray for a short period of time, no later than noon, because fungus spores do not require much time to develop at subtropical temperatures. As a result, leaves might dry out before dusk and spores can be damaged by sunlight before they can germinate. The completely automatic mist propagator uses spray watering to help cuttings take root. Here, a sporadic mist lowers the average air temperature and its daily variance, maintains an almost saturated atmosphere, and lowers the surface temperature of leaves by coating them with a thin layer of water. The rooting medium's temperature drop in temperate locations could be an unfavorable side effect.

The diurnal variation of soil temperatures is reduced by irrigation on well-drained soils, which are the only soil types suitable for irrigation. Irrigation may temporarily lower the soil's average daytime temperature to some degree, depending on the temperature of the water delivered, and may raise the soil's average nighttime temperature. Improved aeration of the subsoil, which results in an increase in oxygen content and a decrease in carbon dioxide content, is one of the most significant effects of draining soil. Drainage is beneficial in this regard even on heavy soils because, despite the fact that it does not significantly expand the air space of the subsurface, it does remove stagnant water from the soil's fissures and fractures and permits oxygenated water to enter the subsoil quickly. Greater root growth, intenser bacterial activity, and the encouragement of oxidation processes, which increase granulation, are further effects of improved aeration. A bare, well-drained soil will have a higher average temperature and a greater depth of diurnal temperature variation than the same soil with poor drainage. Additionally, the soil that has been drained will be able to store more rainwater or irrigation water (and consequently suffer less from erosion caused by surface runoff). Increased soil resistance to farm machinery compression could be a result of draining, which is a highly significant side effect. The need to minimize, if not completely avoid, the use of mechanical vehicles either for cultivation and sowing at the beginning of the crop year or for harvesting and post-harvest activity at the end of the season, severely restricts the cropping program in many areas, such as maritime North-West Europe. In livestock husbandry, excessive animal trampling can also obliterate the structure of soils with an abundance of water. Of course, draining increases soil leaching.

Frost and Frost Management

There are numerous hypotheses and standards for determining when low temperatures harm plants. Because plants can endure temperatures as low as -20°C while dormant in the winter, the issue is even more challenging. However, after the plant has started growing, temperatures a few degrees below freezing point may be lethal in several stages of the plant's development, especially when it is flowering. The development of ice crystals both within and outside of plant cells is currently the most widely accepted idea for the harm and death brought on by frost. The cell sap is a fluid that freezes at a temperature just below zero degrees Celsius. However, water in the gaps between cells may be pure or almost pure, in which case it only freezes at temperatures just below 0°C. Therefore, extra-cellular ice production is more frequent. In this context, the specifics of biological processes like damaged protoplasm, cell structure, etc. are less important than the fact that the phenomena exists. Frost hardiness is a process that involves the plant slowly constructing defenses, and this takes time. Naturally, there won't be a single essential temperature that applies to all kinds of flora. In real life, the combination of the low temperature and the duration of the low temperature determines the harm caused by it.

Prolonged cold temperatures have the potential to harm even the bark of mature forest trees. Although the term "killing frost" is frequently used, it lacks any generic meaning because the extent of the damage varies not only on the temperature but also on the vegetation's recent history. Damage may be larger when cooling comes quickly. There are several different meteorological reasons why below-freezing temperatures can occur during the growing season of crops. We must also take latitude into account in addition to high altitude, which can result in frost even in the tropics. Large-scale subsidence predominates in subtropical climes, say, about 30° latitude, with typically feeble winds. Strong, clear nocturnal inversions can cause significant radiation heat losses at the surface when it is relatively chilly outside, like in the spring. Due to the possibility of massive blocking highs surrounding the break-up of the winter circulation and the potential for short cold spells in a constant westerly circulation, the same radiation frost weather may occur at temperate latitudes, such as about 50° latitude. However, advective frost, which is brought on by arctic air intrusions, is also a possibility at temperate latitudes. Another important factor is the unpredictability of day length at even higher latitudes. The continentality of a region must also be taken into account in climatological study to evaluate the probability of frost at all latitudes.

Protection Against Frost

Both active and passive techniques can be used to guard against frost. The risk that radiation frosts pose to agriculture does not have to be accepted as inevitable. By using wise soil and plant management techniques and, in cases of intense frost, effective action, damage can be avoided or minimized in locations where both the frequency and intensity of such frosts are restricted.

Passive Methods

The focus of passive approaches isn't so much on combating frost as it is on finding ways to greatly lessen the harm that occurs during the frost phase. These techniques involve safeguards taken at the time of planting and are used far before there is a direct threat of frost. Before planting is the greatest time to shield a crop from frost damage. The two primary passive frost-prevention strategies are wise soil and crop management and correct site selection. The third preventive measure—plant selection and breeding of hardier winter wheat and oat types, as well as more frost-resistant strawberry cultivars—is crucial but is outside the purview of this book. The same can be said for agricultural practices like trimming to postpone flowering or dousing plants with pesticides and hormones to increase their resistance to frost. Radiation frosts and advective frosts are the two forms of frosts, and the former has received practically all of the focus in frost studies. There is a widespread belief that the advection of cool arctic air cannot be prevented. However, the likelihood of advective frost should still be taken into account while choosing a site. If a climatological study shows that particular wind directions are linked to the most harmfully low air temperatures during the frost-sensitive seasons, it is necessary to take cover from those winds. Due to the ability of big water bodies to moderate climatic extremes, pure advective frosts are diminished there. On the other hand, radiation frosts can still destroy cranberries on the dunes that are a few hundred meters from the sea. The benefit of being close to things like lakes or warm cities should not be disregarded, either. When paired with cold advection, even modest radiation frosts might result in extremely detrimental effects.

Following this discussion of soil and crop management, it can now be concluded that it is wise to optimize the soil conditions in order for the soil heat flow to contribute as much as possible to the energy balance during frosty nights. Since the soil must be compact and not isolated from the air above it in order to function as a heat source, this may need a variety of procedures, some of which may be long-term. Instead of in the spring, plowing in the autumn allows the soil to become reasonably compact throughout the course of the winter. Before the start of crop flowering, winter mulch should be removed. Mechanical weeding can significantly loosen soil, but chemical weedkillers should also be taken into consideration if springtime weeding is necessary. It is best to utilize tools like rollers to compact the soil if it is too loose for whatever reason. Irrigating soil before a potential frost is an intriguing technique to boost its capacity and thermal conductivity. This is especially effective in the fall because the soil is still very warm from the summer at that time of year. For this, furrow irrigation works fairly well. Flooding, a more extreme practice, has also been utilized. Possible drawbacks include the potential for waterlogging of the soil, which could harm plant roots and soil structure, as well as a potential slowdown in the temperature rise during the daylight after a frost. Crop planting needs to be planned out optimally in addition to managing the soil to reduce frost. They should be grown to maximize cold air drainage, especially on slopes. If at all possible, the planting date should be selected climatologically to avoid the peak frost risk period, when flowering is most likely to occur.

Active Methods

Active frost protection techniques can be rather expensive. The worth of the crop determines how they are used.

Thermal insulation: If possible, crop covering roofs should be made of insulating materials; a straw mat is a good example. This way, the top of the cover will be cooled during frost instead of the crop below. A "microclimate" with a significant input of soil heat flux should be able to develop beneath the roof, and this microclimate requires some ventilation. An efficient approach to kill a plant during a frost is to tie a tiny plastic bag over it. When using small structures like paper shields, protective cones, etc., these variables must be taken into account. A novel approach is to cover the crop with a non-toxic protein foam. The technique might benefit from being automated (so avoiding the manual labor common to many other methods). If subsequent frigid nights are forecast, the foam can stay on the crops for a while - provided, of course, that no associated plant damage occurs.

Interception of outgoing radiation: By generating man-made smoke screens or water mist clouds, radiation can be blocked. Since the particles must be able to capture long-wave radiation from the ground, the effectiveness of such a cloud cannot be assessed by its apparent density. Theoretically, water droplets in a man-made cloud should be more effective than smoke screen particles, however, mist production methods are still being developed. Only weak radiation frosts are known to have significant thermal impacts (of a few degrees). Additionally, as the wind blows the clouds away, they must be continuously formed. This method is only sometimes used because of the pollution that continuous smoke creation causes.

Air mixing: Cloudless radiation evenings cause a chilly layer of air to build close to the surface as the air in contact with the ground cools. The air layer at about 50 meters may be noticeably warmer than the surface layer of air as the temperature rises from the surface upward. The opposite of the typical daylight scenario in which temperature drops with height is this, which is referred to as a temperature inversion. The temperature of the air layer close to the ground can be increased by mixing the warmer upper air above the inversion with the colder layers below. Utilizing wind generators, fans, and even helicopters has been accomplished. The effectiveness of this method of fighting frost relies on how strong the inversion is. Additionally, it has been demonstrated that for gusts greater than 2 m s-1, natural mixing is already so powerful that a wind machine isn't much more useful. Effectively large wind turbines are expensive. A study of inversion strengths on frost nights is a required step before implementing air mixing. Unfortunately, because typical radiosondes don't provide usable information on surface layer temperatures, climatological statistics on inversions are extremely rare and challenging to produce. Wind machines have been employed to protect the delicate citrus crop in California, a region with very strong inversions and weak frosts, but this does not necessarily imply that the technique will be effective in other climates.

Sprinkling: The benefits of using sprinklers or overhead irrigation to prevent frost are clear, and they warrant careful consideration. This procedure needs a lot of water, the right irrigation equipment, and well-drained soil. It only applies if the plant is able to withstand the ice burden that develops on branches, twigs, and leaves. This approach makes up for heat loss through radiation by using latent heat, which is generated when water freezes. Enough energy is generated when one gram of water freezes to raise 80 grams' temperature by one degree. When a tiny layer of water covers a leaf or fruit bud, the heat released when the water freezes is enough to keep that area of the plant's temperature above 0°C. Even though a layer of ice is rapidly forming and the ambient temperature drops several degrees below freezing, as long as this film of water is kept in place consistently, the temperature of the plant will remain above freezing. Sprinkling needs to start early enough to be efficient. As long as the critical temperatures for plant tissues are present, which is typically until the air temperature has risen well above zero and ice starts to fall from the plants, starting the safe practice when the wet-bulb temperature approaches freezing point and continuing it throughout the duration of the practice. For low-growing plants, the recommended minimum sprinkling rates are 1.5 to 2.0 mm h-1; 2.0 mm h-1 for fruit trees; and 2.0 to 2.5 mm h-1 for vines. These rates must be raised in low relative humidity situations to account for additional loss and cooling through evaporation. On the other hand, It is important to apply water at the slowest possible rate. A sprinkler installation has the financial benefit of serving many additional purposes at the farm, including irrigation and pest control.

Direct air and plant heating: The provision of radiation from adequate heaters or small flames is a traditional means of mitigating frost damage. In many parts of the world, heating is a common technique in highly-valued orchards. In this way, the temperature in the orchard or field may be maintained above the critical temperature for the specific crop. The most popular fuel for heating is diesel oil, which is utilized more frequently than solid fuels, gas, and various types of waste. The actual advantage to the plant is partially attributable to the rise in air temperature, but also, and significantly, to the heat that the heaters radiate. Oil and gas heaters heat the air mostly through convection, but depending on the model, 10–30% of the heat that is sent into the orchard may also be radiated by the burners' hot surfaces. The area to be covered should have heaters placed pretty evenly across it, although it is best to place more heaters along the upwind edge and in the chillier areas. Usually, heaters need to be spaced 5 meters apart. The air can be heated more effectively by many little fires than by a few huge ones. Large flames may produce columns of air that rise quickly and break through the inversion ceiling without spreading, allowing the normal circulation of warmer and cooler air to mix. Large fires might cause more harm than help since they rapidly pull in chilly air while also disturbing the ceiling layer. Only shallow inversion layers, less than 10 m high, and light frost appear to be conducive to heating. The average amount of fuel consumed per acre each hour is 300 liters of diesel oil (see WMO-TN No. 157). Frost fighting with heating is typically not cost-effective if harmful frosts occur more than once or twice every season.

Topo-climate research of the area should be the first step in any strategy and plan to combat the threat of frost. Only sprinkling and thermal insulation may be helpful if hard frosts frequently occur during seasons when the crop is in a vulnerable stage of growth. The other approaches may also be sufficiently cost-effective in locations with just light frosts to warrant consideration. The sole agrometeorological forecast provided in almost every country is frost forecasting. The oldest and most basic forecasting technique is to project evening wet-bulb temperature observations as predicted minimum temperatures, frequently with a locally determined constant adjustment of a few degrees (Kammerman's rule) and occasional synoptic corrections. This strategy can take many different forms, such as issuing frost warnings when the dew point drops below 2°C. Regression analysis is frequently used to determine the value of other pertinent characteristics (cloudiness, or length of night). A second strategy is to calculate the surface's radiational cooling, typically using Brunt's formula for long-wave irradiation (there are operationally more reliable estimation formulae now available). Advection was overlooked by early models, however, this has since been fixed. However, very little validation of such surface minimum temperature predictions has been done. Lack of verification data is a problem since, for instance, there are concerns about the representativity of the minimum temperature of grass. The traditional method of forecasting involves estimating the minimum temperature that should be expected at shelter height over a reference surface, adding subjective modifications for surface or orchard conditions, and finally providing the prediction.

Artificial Climate in Glasshouses and Stables

Glasshouses

A glasshouse serves two complementing purposes: it acts as a sort of weatherproofing by blocking the wind from the outside and preventing precipitation from reaching plants. In addition, it tries to increase ambient and soil temperatures, lessen the likelihood of damage from low temperatures, and offer capabilities for controlling soil moisture and air humidity. Equipment can be added to complement natural lighting, influencing the amount of light and its duration, or equipment to change the amount of humidity and carbon dioxide in the air. Normal glass transmits solar radiation's short-wavelength rays but absorbs its long-wavelength (heat) rays. These results support the frequently cited "greenhouse effect," which holds that glass works as a one-way conduit for radiation. Therefore, it must be assumed that the classic "greenhouse effect" is only partially to blame for the reduced heat loss and subsequent maintenance of temperature in the glasshouse.

The environment inside a glasshouse is similar to that of a desert. Since there is no precipitation, an empty "Dutch light" glasshouse can get as hot as solar boilers at 80°C due to a lack of evaporative cooling. The lack of convective heat loss, rather than the greenhouse effect, is the cause of this; hot air is contained in a more-or-less airtight box, from which heat is transported to the ground and the structure. A glasshouse loses heat through convection, radiation, and other methods in addition to the inevitable leaks that occur through any construction, no matter how carefully built. Glass allows for the transfer of heat, which is then dissipated from the outside by convection and radiation. Glass transmits poorly when compared to metals, but it does so better than many insulating and construction materials. The discovery that a glasshouse's night temperature can be maintained at a greater level if mats intended to offer shade during the day are rolled out at night demonstrates that conduction through the glass is not negligible. The structure and its contents, as well as the surface layers, absorb solar radiation. Heat is kept and the amount stored rises above what happens in the open in the absence of convective exchange. The size of the glasshouse will affect temperature variations.

The soil in the greenhouse is heated quite uniformly in the summer. Depending on the size of the greenhouse, there may be a horizontal fall-off in the upper soil layer's temperature from the middle to the edge of the greenhouse during the winter when outside temperatures are low of the order of 1°C m^{-1}. The difference between the temperature inside and outside during the day should be bigger on bright days than on cloudy days. A few hours after sunrise, soil temperatures at 0.1 m below the bare soil surface are also advantageous. However, the earlier growth encouraged under glass will be exposed to spring frosts, thus the effectiveness of protective mechanisms against frost is quite important practically. When the outside temperature dropped below 1°C, additional tests with movable covers included the additional treatment of spritzing the glass' outside with water. Though with little irradiance and high levels of transpiration the leaf temperature may fall below that of the environment, plant leaf temperature typically runs parallel to and somewhat above air temperature (i.e. the leaf is acting as a wet bulb).

In non-ventilated greenhouses, changes in absolute humidity are closely correlated with soil temperature. Humidity increases with soil warming (increased evaporation) and decreases with soil cooling. Even though the evaporation rate is raised and the soil temperature is maintained constant, the absolute humidity is decreased by increasing the air exchange (ventilation). An equilibrium forms between the greenhouse's supply of water vapor and the absolute humidity in the open air while the temperature is remained constant (due to ventilation). Both the absolute and relative humidity levels in the greenhouse are at this moment constant. Therefore, the greenhouse can be made to have dry air (with low relative humidity) by either increasing the air temperature or by using ventilation. In a closed greenhouse with a low air temperature, moist air can be produced by boosting soil evaporation. If the soil is sufficiently moist, then radiant or soil heating devices are typically needed for this. However, because the earth is typically kept a little bit colder in these situations, air heating typically results in drier air than other heating methods. Since air movement is constrained and hence has a limited impact, the saturation deficit of the air dominates the evapotranspiration process under glass. The deficiency is determined by the temperature of the surfaces and plants, which is controlled by the amount of sunlight. The rate of transpiration increases along with solar radiation intensity, which in turn causes the plant to use more water. If the greenhouse is ventilated during sunny weather, dry external air lowers the vapor concentration in its atmosphere, increasing the saturation deficit. The moist soil surface evapotranspires more, but plants can only produce so much transpiration at one time. The stomata of the plant close, limiting the transport of water, when the plant's water balance reaches a certain point. Periods of high sun irradiation are intimately related to transpiration and radiation. Both have very equal absolute values when stated in the same units, meaning that transpiration makes use of all radiation. The two move simultaneously at lesser intensities, but transpiration only contributes a small portion of the radiant input.

For photosynthesis to occur, carbon dioxide is essential, hence it is important to make sure there is a sufficient supply to suit the plant's needs. The plants' requirement for CO_2 rises in the improved growing conditions generated in the glasshouse (particularly if additional radiant energy of the right wavelength is supplied by lamps). However, due to the glasshouse's isolation from external sources, the CO_2 content there may decrease. As a result, it becomes desirable to artificially add CO_2 to a glasshouse to satisfy the plant requirements. Practically speaking, even when the glasshouse is surrounded by other glasshouses that emit CO_2, a CO_2 concentration more than roughly 600 ppm cannot be sustained for any length of time under normal conditions of airtightness of the building and ventilation. It is common to believe that because glasshouse users create their own climate, they don't require meteorological data. This is not completely accurate. For starters, a glasshouse has a limited capacity for heat, and it typically takes an hour to change the heating or ventilation. As a result, glasshouses are extremely vulnerable to an abrupt rise in radiation. In the summer, unventilated temperatures can reach over 70°C. Information on current and anticipated humidity levels is a crucial necessity as well, allowing for rapid climate control modifications. There is a risk of significant maladjustment when such management is based on local data without any prediction information.

Livestock Stables

It is instructive to quickly explore another significant agricultural artificial climate, namely the atmosphere in animal stables, in addition to the climate in glasshouses. We attempt to build artificial climates that are most beneficial to crops in a glasshouse (or its less permanent analog, the plastic tunnel). We make an effort to keep the animals at ease at the stables. Ventilation is crucial for comfort in a number of ways. Sweating becomes challenging in hot, muggy conditions. The issue of removing extra heat and moisture produced by animal breath is exacerbated in stables during windless and humid anticyclonic weather, yet in such conditions, the temperature drops quickly at night. Animal health is negatively impacted by sporadic cold drafts caused by air sensors inside the stable. Extra heat loss is undesirable in cold weather, and a strong breeze causes wind chill.

CHAPTER

8 Agrometeorological Observatory

CONTENTS

OVERVIEW

An agro-meteorological observatory is a location where all the equipment is kept in order to monitor and record various weather elements and parameters at predetermined intervals. Reliable crop-weather relations can be found once the data have been collected for a long enough period of time and have undergone statistical analysis. Consequently, it is crucial to record weather occurrences at an agro-meteorological observatory in order to better understand crop-weather interactions and increase crop productivity. The observatory site selection should adhere to the following fundamental criteria in order to guarantee that the observations are accurate representations of the location and comparable to those produced at other observatories:

- The location needs to be a good representation of the neighborhood.
- As much as feasible, the observatory should be in the middle of the farm.
- During the rainy season, the observatory should not be flooded and should be freely accessible.
- Away from irrigation channels and tall objects like mountains, trees, and buildings. If these structures do exist, the site should be 10 times as far away as their height.
- The area should be level, with no sloping ground.
- The location ought to be chosen once and for all.

A well-exposed, leveled plot about 60 m (N-S) x 40 m (E-W) and roughly in the middle of the farm would make the perfect location for an observatory. To prevent trespassing, the observatory plot should be surrounded by a 1.2 m-high barbed wire fence. The barbed wire fence needs to have a gate in the proper place. For example, in India, under the technical direction of IMD, Pune, agro-meteorological observatories are most frequently built up in all Agricultural Universities.

CLASSIFICATION OF AGRO-METEOROLOGICAL OBSERVATORY

The purpose of a climatological observatory is to examine the impact of climate. A synoptic observatory is kept up for weather research. For agricultural research, an agro-meteorological observatory is maintained.

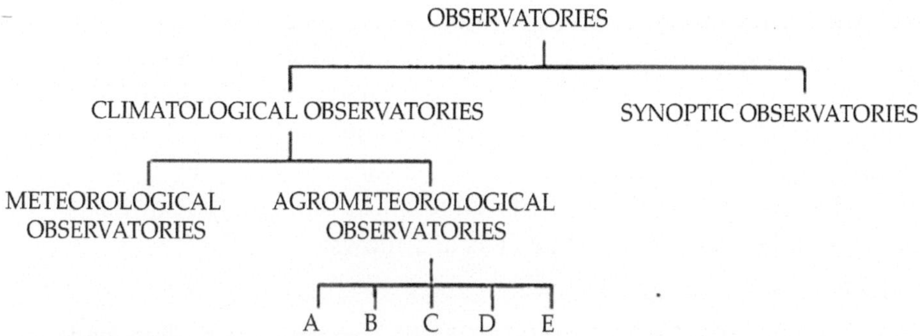

***Figure 44:** Agro-meteorological Observatory classification.*

Agro-meteorological observatories are divided into three major (ABC) and two minor (DE) groups based on the instruments that are available and the quantity of weather elements that are observed (classes). Observatories of Class A, B, and C are equipped with both eye reading and self-recording gear. Three hours apart throughout the day are set out for Class A observatories, three times daily for Class B, and once daily for Class C.

A-Type Observatory

The A-Type or Principal Agro-meteorological Observatory is located at Pune (Maharashtra). Its main activities include:

- recording a vast number of agriculturally relevant meteorological factors (routine work).
- work being done experimentally at the main meteorological observatory.
- coordinating the publication of monthly crop outlooks, weather services for farmers, and the coordination of all India coordinated crop-weather projects on various crops.
- providing state agriculture departments with technical support.
- creation of India's agro-climatic atlases and evaporation measurement.
- publishing agricultural weather data and infographics.
- works in conjunction with WMO in the fields of agrometeorology and related fields.

Most data is continuously recorded by self-recording instruments. These are backed up and reinforced by spot observations made at 0700 and 1400 local mean time (LMT) and at 0830 and 1730 IST during synoptic hours.

Calculation of Local Mean Time: The surface of the earth is divided into 24 time zones, the way in which there are 24 hours in a day. The time established in each zone is 'standard time'. The Indian Standard Time (IST) is the Local Mean Time (LMT) for the longitude of 82°50′ E, which passes through Allahabad. The LMT is based on the transit of the mean sun. To calculate LMT from IST, it is essential to know the longitude of the station. The relation between IST and LMT is as follows:

$$LMT = IST - 4\ (\lambda s - \lambda)$$

where, λS= standard longitude (82°50′ E) passing over Allahabad

λ = longitude of the station for which local time is calculated.

B-Type Observatories

On a regular basis, meteorological and biological information is provided by the B-type observatories, also known as Ordinary Agro-meteorological Observatories.

INSTRUMENTS INSTALLED IN OBSERVATORY

Table 22: *Meteorological Instruments*

Instrument	Parameter Measured
Standard Rain gauge	Rainfall
Automatic Rain gauge	Continuous record of rainfall, storm
Aneroid/Mercury	Barometer Atmospheric Pressure
Cup - counter Anemometer	Wind - run, Wind Speed
Campbell - Stokes	Sunshine Recorder Sunshine hours
Evaporation Pan	Evaporation
Stephenson Screen	Housing for instruments
Dry & Wet bulb Thermometer	Dry & Wet bulb temperature
Thermo - hygrograph	Temperature & Humidity
Wind Vane	Wind Direction

Rain Gauge

A rain gauge, also called udometer, pluviametior, pluviometer, ombrometer, and hyetometer, is a tool that hydrologists and meteorologists use to collect and measure the amount of liquid precipitation over a certain region over time. It is used to measure rainfall amount as well as the depth of precipitation (often measured in mm) that falls over a unit area. An 8-inch Standard Rain gauge (SRG) is essentially a container with an open mouth and straight sides. SRG is required for every climatological station in the "A" Network. The 11-inch plastic gauge can be used in stations other than those in the "A" Network even if the SRG is preferred. The 8-inch conventional rain gauge is a straightforward, non-recording instrument with four main parts. These consist of:

- Measuring Stick
- Overflow Can
- Collector Funnel
- Measuring Tube

Figure 45: *Rain gauge with digital data logger correctly working at Perugia (central Italy).*

Source: Morbidelli R, Saltalippi C, Dari J, Flammini A. A Review on Rainfall Data Resolution and Its Role in the Hydrological Practice. Water. 2021; 13(8):1012. https://doi.org/10.3390/w13081012

The standard gauge, tipping bucket gauge, and weighing gauge are the three main varieties of rain gauges. Although the fundamental operation of rain gauges often does not differ from these main categories of rain gauges, additional distinguishing characteristics can be established, such as how they are configured and how they deliver data. By catching and gathering rainwater into the funnel, the rain gauge functions. Rainwater is emptied from the funnel into the cup below. Additionally, the funnel's design aids in keeping out foreign objects like leaves. The liquid in the cylinder-shaped cup represents the total amount of rainfall. An observer can determine the precise amount of rain that fell after the storm has passed. A meteorologist can determine which locations experienced the highest rainfall using data compiled from a number of rain gauges dispersed throughout the area. Drought monitoring can also be done using the data. One of the most crucial applications for rain gauges worldwide, particularly when it comes to agriculture, is the monitoring of droughts. A tipping bucket rain gauge operates very differently from a traditional gauge. One of two little buckets is reached by the receiving funnel. One bucket can be filled in one hundredth of an inch. The liquid "tipps" into the gauge's outer shell as a result, prompting the second bucket to take its position. The procedure is then repeated. This gauge, which enables accurate monitoring of rainfall intensity and volume, has become commonplace for wireless weather stations. Every time a bucket tips, an electric contact is created, causing a pen to register a mark, according to "Essentials of Meteorology." Wireless digital tipping bucket gauges are widely used now, although they continue to employ the same fundamental technology. The universal weighing rain gauge is the best choice for climatology application, according to the Albany, New York, National Weather Service office. This is due to a vacuum that takes wind effects into account and lets more rain into the gauge. Because the weighing mechanism at the bottom of the collector may be used to measure depth and time at the same time, these gauges are particularly accurate at detecting rainfall intensity. Much to the older tipping bucket gauges, recording is done in the same manner.

Cylinder Ombrometer: The most basic ombrometer is just a cylinder-shaped measuring device with a funnel on the end. Rain that has been gathered by the funnel enters the measuring cylinder where it can be measured against a scale. The 8-inch standard rain gauge, which has been employed by the American National Weather Service for more than 100 years, is the most common variety of cylinder ombrometer. The cylinder needs to be regularly emptied, which is this sort of rain gauge's main drawback.

Tipping Bucket Ombrometer: The funnel portion of the tipping bucket ombrometer empties into a bucket. A second bucket is placed to catch falling rain after the first bucket falls over due to a critical volume. A computer receives an electrical signal when a bucket overturns. This makes it possible to record both the rate and total amount of rainfall.

Optical Ombrometer: A laser beam is directed towards an optical detector that is only a short distance away as part of an optical ombrometer. Rain is able to travel between the laser and the detector through holes in the device's enclosure. The optical detector signal is reduced when this happens. This technology offers far higher resolution than traditional ombrometers since it can detect individual raindrops as they fall. Because optical ombrometers can be built using tiny solid-state lasers and detectors, they can be applied to tasks other than just measuring rainfall. For instance, they can be applied to car windshields as rain sensors.

Weighing Ombrometer: A collection cylinder rests atop a pair of digital scales in a weighing ombrometer. The weight increases as the water builds up in the cylinder, and a computer receives this information. The density of the water and the container's measurements are then used to determine how much rain has gathered inside the collection cylinder.

Barometer

A barometer is a device that measures the variation in air pressure with altitude, either above or below sea level. In the year 1643, an Italian scientist named Evangelista Torricelli created the barometer. He investigated the properties of mercury in a vacuum and developed the first barometer based on this research. In order for a barometer to function, the mercury weight inside the glass tube must be balanced with respect to the atmospheric pressure.

The mercury level will rise if the mercury's weight is less than the air pressure. Additionally, the mercury level will drop if the weight of the mercury is greater than the atmospheric pressure. The inch scale marked on the glass tube is used to measure the mercury's rise and fall. A barometer can be classified into two categories:

Mercury Barometer: The most often used barometer is this one. This was created in 1643 by Torricelli. It is made up of a glass column with markings for inches and mercury. The tube's bottom end is resting in a mercury-filled cup known as a cistern, and the top end is closed. These barometers are adjusted for the local gravity value and the surrounding temperature in order to increase accuracy. Due to its density, mercury is employed for several purposes.

Aneroid Barometer: This barometer is built of solid materials. It is made up of an aneroid capsule, a thin, pliable metal container constructed of a copper-beryllium alloy. The box is hermetically sealed to improve the lever and spring expansion and contraction caused by changes in atmospheric pressure outside the box.

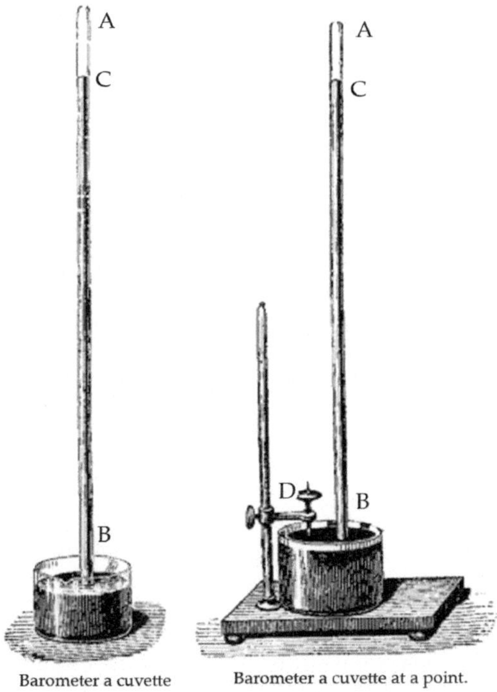

***Figure 46**: Torricellian barometer.*

Source: Public Domain, https://commons.wikimedia.org/w/index.php?curid=295943

Cup - counter Anemometer

The Cup Counter Anemometer is a passive device that uses a mechanical counter to indicate the complete run of wind past the observation location. The mechanical counter's observation glass is angled to make it simple to read from below. The average wind speed for any period of interest can be determined by noting the counter reading at the start and end of that time. To prevent corrosion and ensure a long working life, the instrument is constructed entirely out of brass, copper, stainless steel, and plastic. A conical shield shields the upper spindle bearing from precipitation. All other bearings are self-lubricating, so the only maintenance needed is an annual lubrication of the top spindle bearing.

(a) **(b)**

Figure 47: *Cup - counter Anemometer.*

Source: Roibas-Millan E, Cubas J, Pindado S. Studies on Cup Anemometer Performances Carried out at IDR/UPM Institute. Past and Present Research. Energies. 2017; 10(11):1860. https://doi.org/10.3390/en10111860

Campbell - Stokes Sunshine Recorder

A particular kind of sunshine recorder is the Campbell-Stokes recorder, often known as the Stokes sphere. John Francis Campbell created it in 1853, and Sir George Gabriel Stokes modified it in 1879. A glass sphere was positioned inside a wooden bowl in Campbell's original design, and the sun left a mark on the bowl. Stokes improved the design by utilizing metal for the housing and placing a card holder behind the sphere. The device is made to keep track of how many hours of direct sunlight it takes for the card to become damaged. Little has changed since this fundamental unit was first used, and it still is. Outside of the US, it is frequently employed (the Marvin sunshine recorder is generally the instrument used by the U.S. National Weather Service).

Figure 48: *Campbell - Stokes Sunshine Recorder.*

Source: By Bidgee, CC BY 3.0, https://commons.wikimedia.org/w/index.php?curid=3910724

The crystal ball, which normally has a diameter of 10 cm (4 inches), is made to direct sunlight onto a card that is affixed on the back and rests on a stand. To account for the sun's height during the year, the card is kept in place by grooves, of which there are three overlapping sets. Each day's recording is put on a separate card. The equinox cards are used in the northern hemisphere from 1 March to 11 April and from 3 September to 14 October. The winter cards are used from 15 October to 29 February. For this reason, the summer cards are utilized from 12 April to 2 September. Each card is labeled with the hour and is read in tenths, with local noon in the middle. To record the most amount of sunlight possible in the northern hemisphere, the device is mounted on a pedestal and pointed south. It is critical to position the device where no flagpoles, buildings, or trees will obstruct the sun. The addition of a second, north-facing sphere and card to the basic polar unit modifies it to record the sunlight during the summer, when it is in the sky for 24 hours.

Evaporation Pan

The impacts of numerous climate factors, including temperature, humidity, rain fall, drought dispersion, sun radiation, and wind, are combined or integrated in the assessment of pan evaporation. On hot, windy, dry, sunny days, evaporation is greatest; on cool, calm, and humid days, evaporation is much reduced. Farmers and ranchers can estimate how much water their crops will require by measuring pan evaporation. During observations to estimate the amount of evaporation at a certain place, water is kept in an evaporation pan. These pans come in a variety of sizes and shapes, with circles and squares being the most popular. The "Class A" evaporation pan and the "Sunken Colorado Pan" are the two pans that are most well-known. A Symon's Pan (or occasionally a Symon's Tank) is employed in Europe, India, and South Africa. Frequently, water level sensors and an adjacent tiny weather station automate the evaporation pans. All across the world, different evaporation pans are employed. There are formulas for converting between various pan types and between measurements that are typical of the environment. Evaporation pans installation procedures have also been studied in order to improve the accuracy and repeatability of measurements.

Class A evaporation pan: The Class A evaporation pan is a cylinder with a diameter of 47.5 in (120.7 cm) and a depth of 10 in that has been used as the standard by the National Weather Service in the United States for all of its measurements (25 cm). To keep animals from drinking from it, the pan is frequently contained by a chain link fence and sits on a perfectly leveled wooden base. The amount of water that evaporates from the pan each day is measured in inches as evaporation. The pan is filled exactly two inches (5 cm) from the top on the measurement day. The amount of water needed to fill the pan exactly two inches from the top is measured after 24 hours.

The evaporation is computed after accounting for any precipitation that occurs throughout the 24-hour period. When evaporation exceeds precipitation, measured amounts of water must be dipped from the pan. When the water surface of a Class A pan is frozen, evaporation cannot be recorded in that pan. Unless it is emptied more than once every 24 hours, the Class A Evaporation Pan is only somewhat useful on days with rainfall occurrences of >30mm (203mm rain gauge). Analysis of the daily rainfall and evaporation readings in areas that frequently experience heavy rainstorms reveals that, almost without exception, the daily evaporation is erroneously higher on days with rainfall greater than 30mm (203mm Rain Gauge) than other days in the same month where conditions more conducive to evaporation prevailed. The Class A Evaporation pan will likely overflow in daily rainfall occurrences of >55mm (203mm rain gauge), which is the most frequent and noticeable fault. The effect of heavy or intense rainfall leading to artificially high daily evaporation totals without visible overflow is less obvious but yet more worrisome.

Sunken Colorado pan: The square, unpainted, galvanized iron Colorado pan is 0.92 m (3 ft) on each side and 0.46 m (18 in) deep. As the name implies, it is covered with soil up to the top 5 cm (2 in) of its rim. Using conversion constants, the evaporation from a Sunken Colorado Pan and a Class A pan can be compared. The pan coefficient is roughly 0.8 on an annual basis.

Symons Pan / Tank: The UK Met Office uses the Symons Pan/Tank as a standard instrument. It is a steel container that is 1.83 m (6 ft) on each side and 0.61 m (2 ft) deep. It is buried in the earth and has an above-ground rim that is painted black on the inside. Conversion factors must be applied because its evaporation rate is lower than that of the Class A pan.

Figure 49: *Pan evaporation.*

Source: By Bidgee - Own work, CC BY 3.0, https://commons.wikimedia.org/w/index.php?curid=3910646 Over many parts of the Earth, it is commonly acknowledged that pan evaporation has been declining for the past 50 years. However, there is still some debate regarding the significance of this downward trend in terms of terrestrial evaporation, and it is still not apparent what this means for the global hydrologic cycle. The disagreement derives from opposing theories that these evaporative changes were caused by global radiative dimming or the synergistic interaction of pan and terrestrial evaporation. Actually, these elements work in concert rather than independently.

Stevenson Screen

A Stevenson screen, also known as an instrument shelter, protects meteorological instruments from rain and direct heat radiation from the outside while yet enabling air to freely circulate around them. It is a component of a typical weather station that houses tools like thermometers (normal, maximum/minimum), hygrometers, psychrometers, dew-cells, barometers, and thermographs. A thermometer shelter, an instrument shelter, a thermos-screen, or a thermometer screen are other names for Stevenson screens. Its function is to produce a uniform atmosphere in which to measure air pressure, dewpoint, humidity, and temperature. In order to reflect incoming solar radiation, it is white in color.

Figure 50: *Stevenson screen.*

Source: Yamamoto K, Togami T, Yamaguchi N, Ninomiya S. Machine Learning-Based Calibration of Low-Cost Air Temperature Sensors Using Environmental Data. Sensors. 2017; 17(6):1290. https://doi.org/10.3390/s17061290

The classic Stevenson screen has two louvered sides and is made of wood in a box shape. However, various materials and shapes, like a pyramid, can be used to build a screen. The standard for thermometer height set by the World Meteorological Organization (WMO) is 1.25 to 2 m (4 ft 1 in and 6 ft 7 in) above the ground. The amount of instruments that will be used will determine the interior screen size. Single screens can be as large as 76.5 by 61 by 59.3 cm (30.1 by 24.0 by 23.3 in), while double screens can be as large as 76.5 by 105 by 59.3 cm (30.1 by 41.3 by 23.3 in). Either a wooden pole or four metal or wooden legs support the object. Originally, the top of the screen was made up of two asbestos boards separated by an air space. For health and safety reasons, these asbestos boards have typically been replaced by laminate.

The entire screen is covered in multiple coats of white paint to reflect sunlight, and it often needs to be coated again every two years. To prevent data degradation due to the effects of ground cover, buildings, and trees, the screen's placement is crucial: Even though unfinished, the WMO 2010 proposals provide a solid foundation.

Environment Canada, for instance, advises placing the screen at least twice as far away from an item as its height, such as 20 m (66 ft) from a tree that is 10 m (33 ft) high. The door of the screen should always face north in the northern hemisphere to shield the thermometers from direct sunlight. In polar regions where there is constant sunlight, the observer must be careful to protect the thermometers from the sun while also avoiding raising the temperature due to body heat.

A ship uses a particular kind of Stevenson screen with an eye bolt on the roof. The apparatus is suspended from above and keeps its vertical position even while the ship is moving.

Dry & Wet bulb Thermometer

To assess the level of humidity in the air, it is crucial to know the temperatures of the Dry Bulb, Wet Bulb, and Dew Point. Knowing just two of these numbers will allow you to estimate the moisture content of the air as well as its sensible and latent energy content (enthalpy).

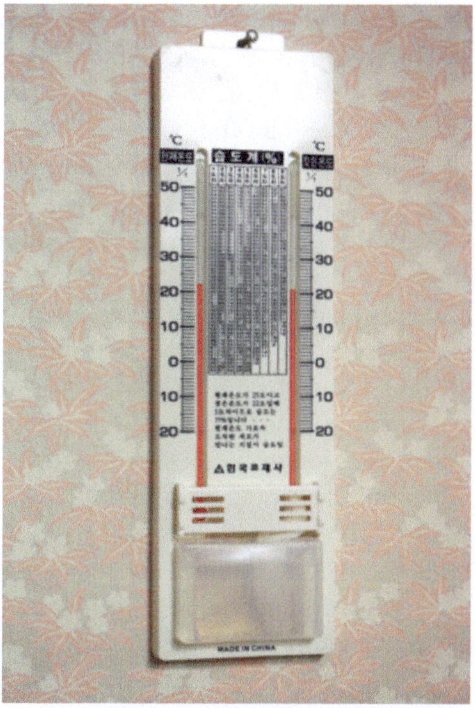

Figure 51: *Dry & Wet bulb Thermometer.*

Source: By Crossmr - Own work, CC BY-SA 3.0, https://commons.wikimedia.org/w/index.php?curid=8718959

Dry Bulb Temperature (T_{db}): The air attribute that is most frequently utilized is the Dry Bulb temperature, sometimes known as "air temperature." Usually, when people talk about the temperature of the air, they mean the temperature of a dry bulb. The ambient air temperature is essentially what the Dry Bulb Temperature refers to. It is known as a "Dry Bulb" because a thermometer used to measure the air temperature does not depend on the humidity of the air. Dry-bulb temperature - T_{db}, can be measured using a normal thermometer freely exposed to the air but shielded from radiation and moisture. The temperature is usually given in degrees Celsius (°C) or degrees Fahrenheit (°F). The SI unit is Kelvin (K). Zero Kelvin equals to -273°C. The dry-bulb temperature, which is displayed along the left side of the Mollier diagram or along the bottom axis of the psychrometric chart, is a measure of the heat content. In the psychrometric chart or the Mollier diagram, constant dry bulb temperatures are represented by vertical or horizontal lines, respectively.

Wet Bulb Temperature (T_{wb}): The adiabatic saturation temperature is the temperature of the wet bulb. With the bulb covered in damp muslin and a thermometer, you may check the temperature of a wet bulb. A "wet bulb temperature" that is lower than the "dry bulb temperature" in the air denotes the adiabatic evaporation of water from the thermometer bulb and the cooling effect. The air's humidity affects both the rate at which the wet bandage on the bulb evaporates and the temperature difference between the wet and dry bulbs. More water vapor in the air causes less of the wet muslin to evaporate. The Dew Point and the Dry Bulb temperature are never higher than the Wet Bulb temperature. There is a dynamic equilibrium between the heat obtained because the wet bulb is cooler than the ambient air and the heat lost due to evaporation for the wet bulb. The temperature of an object that can be reached through evaporative cooling, assuming proper airflow and that the surrounding air temperature is constant, is known as the wet bulb temperature. The condition of the humid air can be determined by combining the dry bulb and wet bulb temperatures in a psychrometric chart or Mollier diagram. The Psychrometric chart is divided into lines of constant wet bulb temperatures that run diagonally from top left to bottom right.

Dew Point Temperature (T_{dp}): The temperature at which water vapor begins to condense out of the air is known as the dew point (the temperature at which air becomes completely saturated). The moisture in the air does not evaporate above this temperature.

- if the dew-point temperature is close to the dry air temperature - the relative humidity is high
- if the dew point is well below the dry air temperature - the relative humidity is low

The dew-point temperature of the air is higher than the temperature in the refrigerator if moisture condenses on a cold bottle removed from the refrigerator. When there is 100 percent relative humidity, the Dew Point temperature will always be the same as the Dry Bulb temperature (the air is at the saturation line). Unless water is added to or withdrawn from the air, the Dew Point tends to remain constant while air temperature varies. By adding some ice cubes and water to a metal can, the Dew Point temperature can be determined. Stir while keeping an eye on the can's exterior. The temperature on the thermometer is quite close to the dew point of the actual air when the vapor in the air begins to condense on the outside of the can.

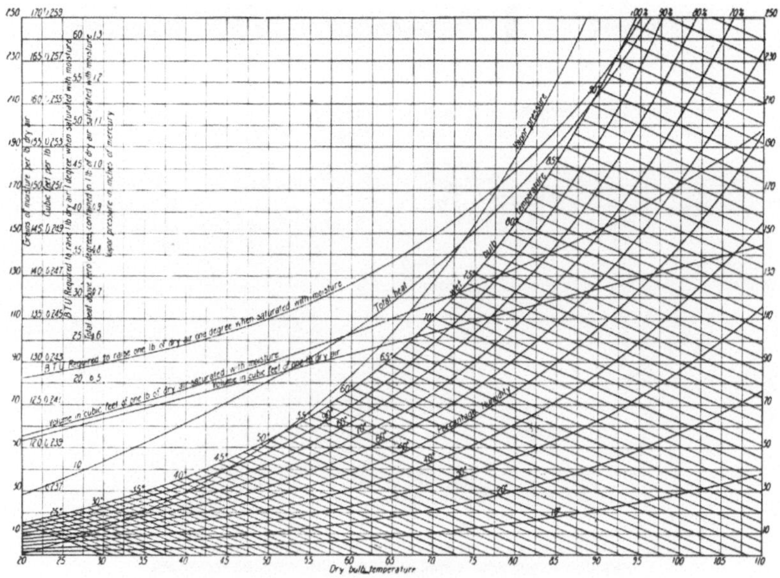

Figure 52: *Psychrometric Chart.*

Source: Teitelbaum E, Miller C, Meggers F. Highway to the Comfort Zone: History of the Psychrometric Chart. Buildings. 2023; 13(3):797. https://doi.org/10.3390/buildings13030797

Thermo - hygrograph

A chart recorder known as a thermo-hygrograph or hygro-thermograph measures and records both temperature and humidity (or dew point). A thermograph for temperature and a hygrograph for humidity are examples of similar instruments that only record one parameter. Several scientists, notably Francis Ronalds, the Honorary Director of the Kew Observatory, described thermographs that record temperature variations using photography as early as 1845. The new UK Met Office established a national observational network in 1867, which Kew Observatory oversaw and deployed an upgraded version of. Then, these instruments were extensively used all over the world. An alternate thermograph design uses a pen to write temperature readings on a rotating cylinder. A bi-metal strip of temperature-sensitive metal that bends as the temperature varies controls a lever that holds the pen at its end. In these machines, humidity can be added using a bundle of human hair.

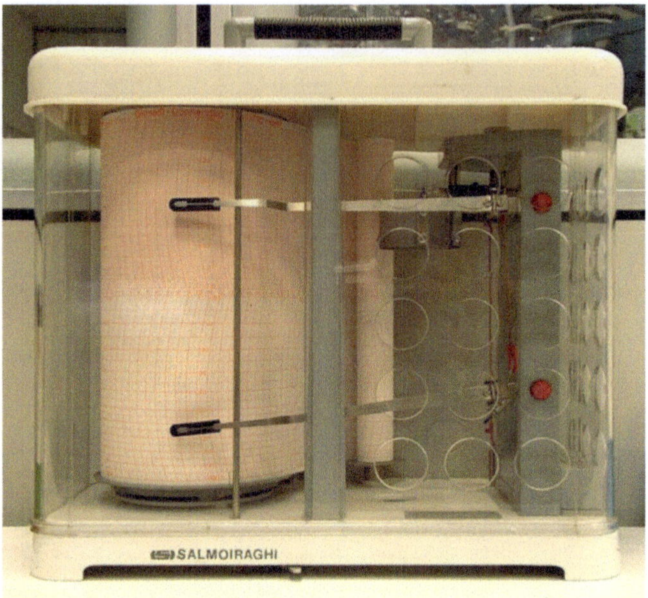

***Figure 53**: Thermo Hygrograph.*

Source: By Luigi Chiesa - Own work, CC BY-SA 3.0, https://commons.wikimedia.org/w/index.php?curid=5106130

Thermo-Hygrographs are devices that use colored traces on a chart to continuously record changes in both the temperature and relative humidity of the atmosphere over time. Thermo-Hygrograph is movable, precise, and simple to use. Variations in relative humidity have a significant impact on human comfort and can have an impact on a variety of manufacturing processes, raw material storage, the weights of goods like wool, paper, tobacco, and confectionery, as well as computer operation. This instrument records both on a single chart since changes in Relative Humidity and Temperature are frequently related. For the Thermo-Hygrograph to capture data effectively and for the cylinder to rotate at the proper pace, constant maintenance is necessary. Additionally, Thermo-Hygrograph calibration must be done on a regular basis. Based on the idea that organic materials swell and contract as the relative humidity varies, thermohygrographs function. The organic substance in this instance is human hair (it has a need to be reconditioned). The hairs are stretched between a fixed pin and a moving pin while being bundled together. The levers that the movable pin is connected to enable the hairs to move more forcefully. The end lever is equipped with a pen that marks the movement of the lever on a chart that is positioned on a rotating cylinder. Relative Humidity can be plotted over a day, a week, or a month by changing the rate of rotation. Additionally, it features a temperature sensor that logs temperature data on the same chart. At the conclusion of each recording period, the chart should be altered, and the pen should be placed on the proper line.

Wind Vane

One of the first meteorological instruments is the wind vane, which shows wind direction. At the top of the rod of the wind vane is a revolving decoration that indicates the direction. There are two globes: a smaller one at the top and a larger one at the bottom of the rod. On the rod between the two globes is a direction pointer with the arrows pointing north, south, east, and west. The vane spins when it is mounted on an elevated shaft or spire, causing its center of pressure to rotate to the leeward and the vane to point into the wind. Degrees from true north are used to determine the wind's direction. The Anglo-Saxon word "fane," which means "flag," is where the term "vane" originates. The direction the wind blows from establishes the wind direction. If the air current is blowing from the West, the wind is westerly.

Figure 54: *Wind vane.*

Source: By Pat (Cletch) Williams from Glen Allen, VA, USA - Romney Ryan RALLY Iin Michigan, CC BY 2.0, https://commons.wikimedia.org/w/index.php?curid=95118278

The directional vane and anemometer are combined in modern aerovanes. Both instruments can use the same axis when they are co-located, and this results in a synchronized readout. types of wind vanes include;

- **Classical style:** It has a fairly straightforward design. It is now used as a decorative element in towers and buildings. The fins in the traditional form are shaped like roasters.
- **Windsock:** These wind vanes can be seen near runways in airports. A windsock can indicate direction, although it typically does not provide precise north, east, west, or south directions. The observer can determine the wind speed based on the inflation of the sock. The pilots will find this to be quite useful during landing and takeoff. Where the direction and strength of the wind are advantageous for the activity, such as a baseball field, golf course, or marine, a windsock can be seen.
- **Digital wind vane:** Simply said, a digital wind vane is more advanced. They are visible in weather stations, which measure the wind speed and direction together with the temperature, pressure, and humidity. It contains an optical sensor that enables the processor to recognize up to 16 positions of direction. On a digital display, this reading is visible.

AGRO-METEOROLOGICAL MODELS AND REMOTE SENSING

Crop modeling technologies, techniques, and approaches have advanced quickly in recent years, similar to many other fields. This evolution has been mostly fueled by technology, from computers to communications. For application at the size of a field, process-oriented models were primarily developed in a research environment. To be useful in practical settings, they need to be greatly simplified. Geostatistical methods, RWGs, and GIS methodologies are now well-known tools. All agrometeorological modeling of weather impacts on crops now includes them. In some way, satellite inputs fall short of their potential and their reputation. Numerous issues, including price, poor user education, and product accessibility, can be used to explain this.

Factors Behind the Inter-Annual Variability of Yields

A time series examination of agricultural statistics reveals that essentially three factors account for the inter-annual fluctuation of crop yields:

i trend,

ii direct weather factors and

iii indirect weather effects, pests, diseases, weed competition, etc.

The fact that weather affects food production at many stages of the food production chain, as seen below, further complicates the subject of weather's impact on food availability.

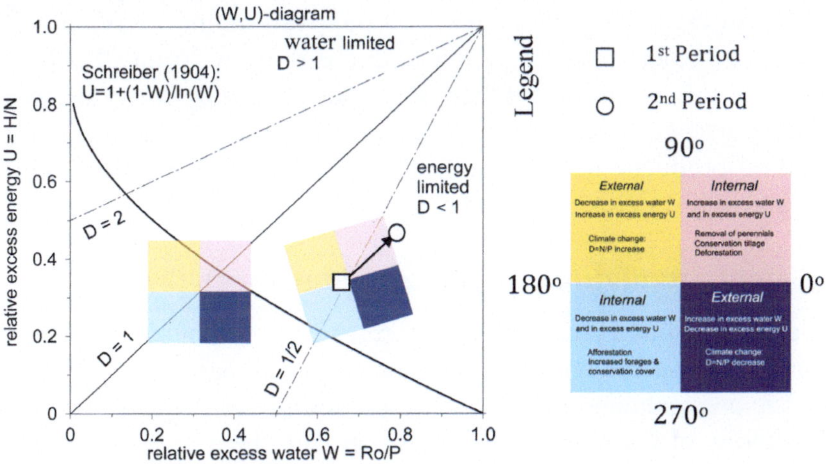

***Figure 55**: Spectrum of the sources of weather-induced variability.*

Source: Cai D, Fraedrich K, Sielmann F, Zhu S, Yu L. Attribution and Causality Analyses of Regional Climate Variability. Land. 2023; 12(4):817. https://doi.org/10.3390/land12040817

As a result, the range of factors that contribute to weather-induced variability is quite broad and includes both direct and indirect causes, such as the availability of water and sunlight, pollination, pests and diseases that affect crops and animals, trafficability, the movement of pathogens by the wind, irrigation, food storage, phenology, etc. The general socio-economic environment has a significant impact on the relative importance of the many causes of variability in food production. The technological component is not very noticeable in many developing nations, and some of the poorest nations have no yield trend at all. In this case, weather conditions might have a significant effect and endanger the food security of millions of people. The same farmers will go through a transition phase where their susceptibility to weather whims will increase as they are increasingly compelled by circumstances to adapt to more commercial farming. The scale used for the assessment has a significant impact on the relative importance of weather variability. The impact of weather is much less noticeable at low levels of aggregation, where a variety of cropping seasons and typologies are combined. Extreme agrometeorological events are factors that frequently occur with high intensities while being unusual (low statistical frequency). Large-scale insect outbreaks, fires, torrential rains, tropical cyclones, etc. are a few examples. They have the potential to cause severe damage to buildings, crops, livestock, fishing equipment, etc., as well as the loss of lives. The impact of severe circumstances is a crucial discovery. Typically, they have significantly less of an impact than the losses caused by recurring weather-related events like droughts, localized pest attacks, the biodegradation of agricultural products, hail, etc., despite the immense harm they can cause locally and at a very specific moment.

Crop Yield Simulation Models and Statistical "Models"

A model is a computer program (and the research that behind it) that simulates (predicts) a plant's behavior (output), such as yield, based on environmental factors (inputs, including management), as well as variables that describe the ecophysiology of the plant (parameters). In fact, there are a lot of different models (as well as associated problems). In actuality, a model is an accounting system that records how much biomass is kept in each organ.

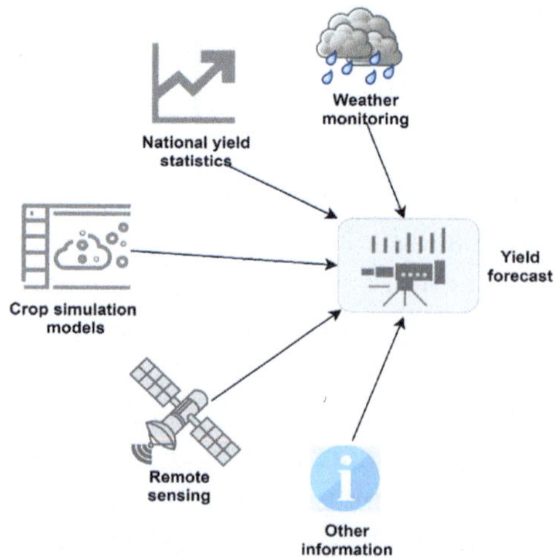

***Figure 56**: Schematic representation of a crop simulation model.*

Source: Batool D, Shahbaz M, Shahzad Asif H, Shaukat K, Alam TM, Hameed IA, Ramzan Z, Waheed A, Aljuaid H, Luo S. A Hybrid Approach to Tea Crop

Yield Prediction Using Simulation Models and Machine Learning. Plants. 2022; 11(15):1925. https://doi.org/10.3390/plants11151925

The model calculates the maximum quantity of plant material (biomass) that may be accumulated each day by comparing factors including leaf area, daylight, water, and nutrient availability (daily growth, to be added to the old biomass). The plant organs are then distributed among the daily growth. The fresh biomass is exposed to the surrounding conditions, etc., the following day.

Models must be calibrated, or confirmed and fine-tuned, by comparison with a variety of characteristics that have really been measured, such as soil moisture, soil nutrient content, actual biomass buildup and yield, etc. Multiple regression equations are statistical "models" that are used to calculate crop yield as a function of one or more agrometeorological factors. Unrealistic values may emerge from computation utilizing raw agrometeorological data, which is a major concern with models. When value-added variables are employed instead of the raw agrometeorological variables, as is typical in crop forecasting systems put up with FAO support, many of the drawbacks of the regression approaches can be avoided. A good example of such a value-added variable is real crop evapotranspiration, which is known to be directly correlated with the amount of solar radiation absorbed by the plant under favorable water supply conditions or mild water stress. Value-added variables are frequently obtained by using straightforward simulation models.

GIS Techniques, Geo-Statistics and Random Weather Generators (RWG)

Although crop projections must be made for wider administrative regions, such as districts, provinces, etc., because models only work with data from a single location (a point), they are based on this data. The groups of methods used to transform point data into broader areas are part of geographic information systems and geo-statistics (statistics of spatial data) (GIS). Finally, the meteorological conditions at the precise area where a model is performed are typically unknown. Although there is no meteorological station nearby, geostatistical tools are now frequently used to estimate the weather conditions at the place where a model is being run. The requirement to predict the anticipated weather between the time of the forecast and harvest arises from the fact that forecasts are made while the crop is still growing in the field. Let's say we are predicting the yield of a crop that was planted in November and is scheduled to be harvested in May. Assume that we are currently at the end of February to get an accurate picture of the weather from November to February. For the "future" weather between March and May, we must make some assumptions. To do this, we need a collection of weather data spanning 30 or 50 years that could plausibly occur. The 30 or 50 yield estimations that result from this can be averaged to provide the current projection. There are many methods, but one that is frequently used is to use Random Weather Generator, a program that uses historical data from the station under investigation to create fictitious weather series for a specific place. Synthetic series, in the words of Göbel (1995), are examples of possible sequences of occurrences rather than predictions of what will occur in the future.

They come in very handy for a variety of uses, from crop forecasts to risk assessment. The stochastic processes (Markov chains) and correlations that exist in the historical data are simulated by RWG. The processes are managed by site-specific coefficients. The ability to generate artificial weather that covers wide areas is made possible by the fact that the coefficients driving a generator can be mapped (spatially interpolated) like any other variable. Note that the fields produced in this way typically lack spatial coherence. Geographic Information Systems are the most commonly associated models and tools. GIS techniques are used to format and show the output and analysis following model runs. They are also used to prepare the geographical input data for the regional applications.

Remote Sensing and Other New Data Sources

Satellite data and simulation models clearly complement one another. The first reason is that remote sensing can help with surface agrometeorological variable estimation. Second, crop modeling is increasingly using inputs from satellites. Despite the current deficiencies of the suggested approaches, there is little doubt that improvements in spatial and spectral resolutions will lead to advancements in the estimation of biomass and the components of the water balance (soil moisture) (LAI and conversion efficiencies).

Early attempts to anticipate crops using satellite data mostly used Vegetation Indices (VI), which are satellite-derived indices related to living green biomass. Although the qualitative application of VIs is now commonplace in many nations, their quantitative application in crop production forecasts has remained unsatisfactory for a variety of reasons.

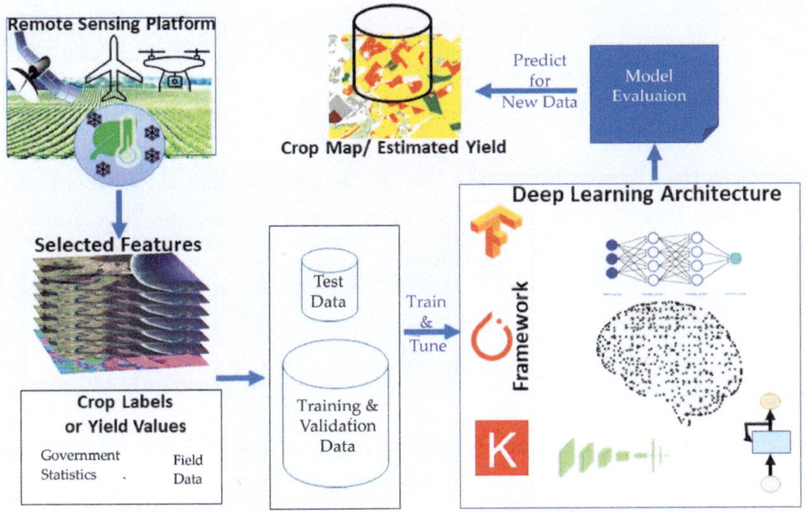

***Figure 57:** Overview of remote sensing in agriculture.*

Source: Joshi A, Pradhan B, Gite S, Chakraborty S. Remote-Sensing Data and Deep-Learning Techniques in Crop Mapping and Yield Prediction: A Systematic Review. Remote Sensing. 2023; 15(8):2014. https://doi.org/10.3390/rs15082014

The use of VIs and other satellite inputs, such as cloud information, as auxiliary variables for stratification and zoning, as well as the area averaging of point data when combined with GIS and geostatistics, are seen to hold some of the biggest potentials for VIs and other satellite inputs. Many times, especially in developing nations, fields are small and irregular in size and shape, crops are frequently combined, etc., so that the sensors essentially measure a mixture of natural vegetation and crops. So it is commonly accepted that crops exhibit patterns of greenness comparable to those of vegetation. In regions with clearly seasonal vegetation, such as semi-arid regions, this is a plausible assumption. At higher spatial resolutions, many of the listed challenges vanish.

Vegetation indices: Since plants reflect a far higher percentage of white sunlight in the green region of the spectrum than in the blue or the red, this fact serves as the foundation for all vegetation indices. In actuality, plants contain different amounts of chlorophylls a and b, which absorb primarily blue light between 0.38 and 0.45 m and red light about 0.675 m. Plants typically appear quite bright when viewed through a near-infrared filter or sensor because they reflect a substantially bigger percentage of near infrared light. For a variety of reasons, things are not so simple. The spectral fingerprints are first obtained experimentally on the ground with known irradiance levels5. Satellite measurements of radiance reveal qualitative (wavelength) and intensity variations in the atmosphere. In addition, the irradiance varies depending on the atmosphere and the angle of incidence (which is to say: the time of the day). Thus, a lot of vegetation indices are "normalized" to at least partially account for the impacts of the atmosphere and the soil, as seen in the well-known Normalized Difference Vegetation Index (NDVI). Naturally, NDVI and other vegetation indices can be calculated at any scale. In fact, US companies have begun offering metric scale, extremely precise NDVI maps of farms.

Surface temperature and evapotranspiration (ETP): The estimation of ETP and surface temperatures using satellite data are related topics that are receiving a lot of attention due to their importance for computations of the soil water balance. The traditional air temperature (Ta) is the temperature of the air measured in a screen, whereas the surface temperature (Ts) is the temperature of the air layer directly in contact with the leaf. The work on Jackson's CWSI is where the interest in the differential between surface and air temperatures first emerged (Crop Water Stress Index; Jackson et. al., 1981). The FAO Water Satisfaction Index (WSI), established by Frère and Popov (1979) as a monitoring tool resulting from a water balance calculation over the entire cycle, is very similar to the concept of the CWSI. Since Ts-Ta is a measurement of the sensible heat flow, which shows how much energy is available for evapotranspiration, it can be demonstrated that, for a given net radiation, there is a direct relationship between real evapotranspiration and Ts-Ta. Significantly less evapotranspiration and stomatal closure are indicated by a substantial difference. A slight variation shows that the water supply is sufficient and that plants do in fact evapotranspire. Low temperature differences are typically a sign of healthy, photosynthesizing crops.

Cold Cloud Duration (CCD): METEOSAT or GMS type geostationary satellites are used to calculate Cold Cloud Duration (CCD). In comparison to NDVI, there are comparatively few atmospheric effects to be addressed in CCD since low cloud-top temperature thresholds equate to high heights. Japanese satellite GMS-4 is in a geostationary orbit at 140E east of the equator. The Visible and Infrared Spin Scan Radiometer (VISSR) imaging sensor, which employs the spin motion of the satellite to scan the planet in an east-west orientation, is affixed to the satellite, which covers much of Asia and the western Pacific Ocean. In addition to four additional scans per day for wind estimation, GMS starts a North-South scan every hour on the half-hour. In the area of food security, CCD has been used frequently to estimate rainfall using a variety of methods. The number of hours the temperature was below a "cold" temperature threshold of roughly -40°C, which corresponds to convective clouds with high vertical extension assumed to cause rainfall, are defined for each pixel for a specified time period (often ten days). The method has only been successfully applied to estimate rainfall in tropical nations. The development of more or less empirical methods is continuing, typically with much better results than with rainfall, despite the fact that the relationship between the solar radiation reaching the ground and clouds is far from direct. This is due, among other things, to the fact that the role of clouds in radiation interception is much more direct than in the production of rainfall. Furthermore, the techniques work in both tropical and temperate countries provided they have been correctly calibrated. The initial strategy involved attempting to calculate rainfall using only CCD and assuming a constant intensity Ri (mm hour-1). The approach is presently being superseded by more or less real-time calibration against ground data, utilizing CCD as an auxiliary variable in the spatial interpolation of rain-gauge measurements, due to the considerable spatial and temporal fluctuation of Ri.

Observations by microwave satellites: There are a number of challenges in using satellites for the direct measurement of surface moisture. Clouds don't typically interfere with active microwave (or radar) satellites operating in the centimetric range of wavelengths, and satellites like ERS-1 and JERS-1 have been taking pictures of the Earth since the early 1990s. Radar is an active sensor because it releases a beam of energy that is then examined after it is reflected back off the surface. This gives it information about the surface, such as the structure of the soil surface or the crop canopy. Active microwave reacts favorably to leaf orientation, row spacing, and even row orientation in crops. Due to the technique's high sensitivity to variables like wind influences on the surface, especially leaf orientation, little operational use has been made of it thus far. The ability to immediately determine soil surface moisture and perhaps crop water content of specific plants is highly anticipated. The centimetric radiation that the surface emits is measured using passive microwave. It is used to calculate the effective temperature, also known as the brightness temperature, which is useful for estimating biomass.

Ground-based weather radar: Similar to microwave satellites, weather radar also functions in the centimeter range. Basically, the method estimates the amount of rain that falls within a 100 km, or occasionally less, circle surrounding the station. Its key benefit is that, compared to other techniques, it may be used to determine the spatial distribution of rain over short time periods with a great deal more accuracy. Similar to satellite rainfall predictions, after calibration against ground data, the best results are obtained over quite long time intervals (days and beyond). According to a source cited by Keane (1998), radar calibrated against two rain gauges spread across 1,000 km2 had the same accuracy as 50 rain gauges during showery circumstances.

Satellite Enhanced Data Interpolation (SEDI): The FAO's Regional Remote Sensing Project in Harare is where the SEDI interpolation technique first appeared. Initially, it was created to interpolate rainfall data gathered at the station level using the extra details offered by METEOSAT CCD. The techniques were effective and adaptable, and they are now frequently used to spatially interpolate other characteristics (e.g., potential evapotranspiration, crop yields, actual crop evapotranspiration estimates, etc.). SEDI makes use of the relationship between an environmental variable, such as NDVI/ biomass and agricultural yields, and the variable to interpolate. Co-kriging, a kriging variant that uses one or more auxiliary variables and takes advantage of both the geographical characteristics of the variable to be interpolated and the correlations between the variable and the auxiliary variables, is one method of doing this. The following three conditions must be met before using the SEDI method:

- The parameter's accessibility as point data for interpolation at various geographic locations (e.g., rainfall, potential evapotranspiration, crop yields). In the current instance, statistical variables were given a coordinate matching to the administrative unit's center of gravity;
- the availability of a background parameter for the same region in the form of a regularly spaced grid (or field) (e.g., the above-mentioned NDVI variables, altitude).
- At least locally, there is a monotone relationship (positive or negative; yield/NDVI is positive, temperature/altitude would be negative). A Spearman rank correlation test can demonstrate the presence and strength of a relationship.

For Early Warning application, the top crop simulation models (CERES, WOFOST, EPIC, and CROPSYST...) are far too complex. Models should concentrate on weather and agricultural factors that are truly accessible in the real world if they are to be utilized frequently in the real world. These models would also be scale-specific, i.e., they would have been created for use over provinces or districts as opposed to individual fields. Both new and updated versions of some outdated utilities are required. Because they are inexpensive and easy to calibrate and use, non-parametric and rule-based models should receive more attention from researchers. Efficiency should also be prioritized over fashion; for example, regression models should still be used in monitoring systems alongside more modern tools like GIS, geo-statistics, and random weather generators. In crop monitoring and forecasting systems, more emphasis needs to be paid to new real-time data. This comprises weather radar and satellite indicators, but it also refers to ground-based direct (and sometimes automatic) measurements of crops and weather, as well as remote sensing by aircraft.

Crop monitoring and forecasting systems should embrace a modular design, and there should be increased tool interoperability. By utilizing building pieces (modules), such as the root component of CERES, the assimilation block of WOFOST, the Leaf-Area Index from the MODIS satellite, etc., users should be able to put together their own monitoring and forecasting systems. The synchronization of data files in the direction of a global self-documented format needs to be the first step. A new class of models should be created to evaluate the impact of extreme weather conditions that physically injure plants (such as frost, sandstorms, and strong winds), as well as a better integration of crops, diseases, and competitors, at different scales. The scale problem is especially important in the context of EW systems because other variables they employ (such as socioeconomic and nutritional data) all have native scales that need to be synchronized. Reduced product costs are part of this. A substantial effort should be made by agrometeorologists and remote sensing specialists (including the satellite-operating agencies) to inform the experts in national agrometeorological services and food security systems about the techniques that are currently available. The accessibility of the new tools and data should be improved in every way at the same time.

REFERENCES

1. Abezghi, T. W. (2003). Estimation of reference evaporation and comparison with ET-gage evaporimeter (Doctoral dissertation).
2. Abu-Hamdeh, N. H. (2003). Thermal properties of soils as affected by density and water content. Biosystems engineering, 86(1), 97-102.
3. Abu-Hamdeh, N. H., & Reeder, R. C. (2000). Soil thermal conductivity effects of density, moisture, salt concentration, and organic matter. Soil science society of America Journal, 64(4), 1285-1290.
4. Abu-Hamdeh, N. H., & Reeder, R. C. (2000). Soil thermal conductivity effects of density, moisture, salt concentration, and organic matter. Soil science society of America Journal, 64(4), 1285-1290.
5. Ahmad, L., Habib Kanth, R., Parvaze, S., Sheraz Mahdi, S., Ahmad, L., Habib Kanth, R., ... &Sheraz Mahdi, S. (2017). Measurement of sunshine duration.
6. Al Nakshabandi, G., &Kohnke, H. (1965). Thermal conductivity and diffusivity of soils as related to moisture tension and other physical properties. Agricultural Meteorology, 2(4), 271-279.
7. Al Nakshabandi, G., &Kohnke, H. (1965). Thermal conductivity and diffusivity of soils as related to moisture tension and other physical properties. Agricultural Meteorology, 2(4), 271-279.
8. Alberto, M. C. R., Wassmann, R., Hirano, T., Miyata, A., Kumar, A., Padre, A., &Amante, M. (2009). CO2/heat fluxes in rice fields: Comparative assessment of flooded and non-flooded fields in the Philippines. Agricultural and Forest Meteorology, 149(10), 1737-1750.

9. Allen, R. G. (1994). An update for the calculation of reference evapotranspiration. IcID Bull., 43(2), 35-92.
10. Anadranistakis, M., Liakatas, A., Alexandris, S., Aggelides, S., Kerkides, P., Rizos, S., &Poulovassilis, A. (1996, September). Soil heat flux in the Penman-Monteith evapotranspiration equation. In II International Symposium on Irrigation of Horticultural Crops 449 (pp. 69-74).
11. Avissar, R., &Mahrer, Y. (1988). Mapping frost-sensitive areas with a three-dimensional local-scale numerical model. Part II: comparison with observations. Journal of Applied Meteorology (1988-2005), 414-426.
12. Azooz, R. H., Lowery, B., Daniel, T. C., & Arshad, M. A. (1997). Impact of tillage and residue management on soil heat flux. Agricultural and forest meteorology, 84(3-4), 207-222.
13. Baier, W. (1969). Concepts of soil moisture availability and their effect on soil moisture estimates from a meteorological budget. Agricultural Meteorology, 6(3), 165-178.
14. Baier, W. (1973). Crop-weather analysis model: review and model development. Journal of Applied Meteorology and Climatology, 12(6), 937-947.
15. Baier, W., & Robertson, G. W. (1966). A new versatile soil moisture budget. Canadian Journal of Plant Science, 46(3), 299-315.
16. Baier, W., & Robertson, G. W. (1968). The performance of soil moisture estimates as compared with the direct use of climatological data for estimating crop yields. Agricultural Meteorology, 5(1), 17-31.
17. Balmer, R., & Balmer, R. T. (2011). Modern engineering thermodynamics-textbook with tables booklet. Academic Press.
18. Barfield, B. J., & Gerber, J. F. (1979). Modification of the aerial environment of plants.
19. Barrett, E. C. (1975). Environmental Survey Satellites and Satellite Data Sources. Geography, 31-39.
20. Baver, L. D., Gardner, W. H., & Gardner, W. R. (1972). The thermal regime of soils. BAVER et al., ed soil physics. New York, John Wiley, 253-280.
21. Bean, A., Alperi, R. W., & Federer, C. A. (1974). A method for categorizing shelterbelt porosity. Agricultural Meteorology, 14(1-2), 417-429.
22. Beljaars, A. C. M. (1987). On the memory of wind standard deviation for upstream roughness. Boundary-layer meteorology, 38(1-2), 95-101.

23. Beljaars, A. C. M., &Holtslag, A. A. M. (1991). Flux parameterization over land surfaces for atmospheric models. Journal of Applied Meteorology and Climatology, 30(3), 327-341.
24. Beljaars, A. C. M., &Holtslag, A. A. M. (1991). Flux parameterization over land surfaces for atmospheric models. Journal of Applied Meteorology and Climatology, 30(3), 327-341.
25. Benichou, P. Le Breton 0., 1987—Prise en compte de la topographie pour la cartographie de champs pluviométriquesstatistiques: la méthodeAurelhy. Agrométéorologie des régions de moyennemontagne. Les colloques de I'INRA, Paris, 51-69.
26. Björkman, O. (1968). Further studies on differentiation of photosynthetic properties in sun and shade ecotypes of Solidago virgaurea. Physiologia plantarum, 21(1), 84-99.
27. Bolz, H. M., & Fritz, H. (1950). Tables and diagrams for the calculation of reflected radiation and re-radiation. Z. Meteorol, 4.
28. Bowen, A. J., & Lindley, D. (1977). A wind-tunnel investigation of the wind speed and turbulence characteristics close to the ground over various escarpment shapes. Boundary-Layer Meteorology, 12, 259-271.
29. Bremer, D. J., & Ham, J. M. (1999). Effect of spring burning on the surface energy balance in a tallgrass prairie. Agricultural and Forest Meteorology, 97(1), 43-54.
30. Bristow, K. L. (1988). The role of mulch and its architecture in modifying soil temperature. Soil Research, 26(2), 269-280.
31. Bristow, K. L. (2002). Thermal conductivity. p. 1209–1226. JH Dane and GC Topp (ed.) Methods of soil analysis. Part 4. SSSA Book Ser. 5. SSSA, Madison, WI. Thermal conductivity. p. 1209–1226. In JH Dane and GC Topp (ed.) Methods of soil analysis. Part 4. SSSA Book Ser. 5. SSSA, Madison, WI.
32. Brutsaert, W. (1982). Evaporation into the atmosphere: Theory. History, and Applications, 1.
33. Brutsaert, W. (2013). Evaporation into the atmosphere: theory, history and applications (Vol. 1). Springer Science & Business Media.
34. Brye, K. R., Norman, J. M., Bundy, L. G., & Gower, S. T. (1999). An equilibrium tension lysimeter for measuring drainage through soil. Soil Science Society of America Journal, 63(3), 536-543.

35. Buchan, G. D. (1989, May). Soil heat flux and soil surface energy balance: a clarification of concepts. In Proceedings of the Fourth Australasian Conference on Heat and Mass Transfer. University of Caterbury, Christchurch, New Zealand.
36. Budyko, M. I. 1963: Atlas teplovogobalansazemnogoshara (atlas of the heat balance of the earth). AkademiyaNauk SSSR.
37. Cahill, A. T., &Parlange, M. B. (1998). On water vapor transport in field soils. Water Resources Research, 34(4), 731-739.
38. Carson, J. E., & Moses, H. (1963). The annual and diurnal heat-exchange cycles in upper layers of soil. Journal of Applied Meteorology and Climatology, 2(3), 397-406.
39. Cellier, P., Richard, G., & Robin, P. (1996). Partition of sensible heat fluxes into bare soil and the atmosphere. Agricultural and forest meteorology, 82(1-4), 245-265.
40. Changnon, S. A. (1992). Contents of climate predictions desired by agricultural decision makers. Journal of Applied Meteorology (1988-2005), 1488-1491.
41. Choudhury, B. J., Idso, S. B., &Reginato, R. J. (1987). Analysis of an empirical model for soil heat flux under a growing wheat crop for estimating evaporation by an infrared-temperature based energy balance equation. Agricultural and Forest Meteorology, 39(4), 283-297.
42. Clothier, B. E., Clawson, K. L., Pinter Jr, P. J., Moran, M. S., Reginato, R. J., & Jackson, R. D. (1986). Estimation of soil heat flux from net radiation during the growth of alfalfa. Agricultural and forest meteorology, 37(4), 319-329.
43. Collier, C. G. (1986). Accuracy of rainfall estimates by radar, Part I: Calibration by telemetering raingauges. Journal of Hydrology, 83(3-4), 207-223.
44. Counihan, J. J. C. R., Hunt, J. C. R., & Jackson, P. S. (1974). Wakes behind two-dimensional surface obstacles in turbulent boundary layers. Journal of Fluid Mechanics, 64(3), 529-564.
45. Crawford, K. C., & Hudson, H. R. (1973). The diurnal wind variation in the lowest 1500 ft in central Oklahoma. June 1966–May 1967. Journal of Applied Meteorology and Climatology, 12(1), 127-132.
46. Çuhac, C., Mäkiranta, A., Välisuo, P., Hiltunen, E., &Elmusrati,

M. (2020). Temperature measurements on a solar and low enthalpy geothermal open-air asphalt surface platform in a cold climate region. Energies, 13(4), 979.

47. Dalton, G. E. (1974). The effect of weather on the choice and operation of harvesting machinery in the United Kingdom. Weather, 29(7), 252-260.
48. Daughtry, C. S. T., Kustas, W. P., Moran, M. S., Pinter Jr, P. J., Jackson, R. D., Brown, P. W., ... & Gay, L. W. (1990). Spectral estimates of net radiation and soil heat flux. Remote sensing of environment, 32(2-3), 111-124.
49. De Bruin, H. A. R., &Keijman, J. Q. (1979). The Priestley-Taylor evaporation model applied to a large, shallow lake in the Netherlands. Journal of Applied Meteorology (1962-1982), 898-903.
50. De Bruin, H. A. R., &Lablans, W. N. (1998). Reference crop evapotranspiration determined with a modified Makkink equation. Hydrological Processes, 12(7), 1053-1062.
51. de Silans, A. P., Monteny, B. A., &Lhomme, J. P. (1997). The correction of soil heat flux measurements to derive an accurate surface energy balance by the Bowen ratio method. Journal of Hydrology, 188, 453-465.
52. De Vries, D. A. (1952). A nonstationary method for determining thermal conductivity of soil in situ. Soil Science, 73(2), 83-90.
53. Deacon, E. L. (1950). The measurement and recording of the heat flux into the soil. Quarterly Journal of the Royal Meteorological Society, 76(330), 479-483.
54. Dean, J. A., &Spaight, J. (1999). Thermodynamic properties: Enthalpies and Gibbs (free) energies of formation, entropies and heat capacities of elements and compounds. Lange's handbook of chemistry, 15th ed. McGraw-Hill, Inc., Burr Ridge, IL.
55. Doorenbos, J. (1977). Crop water requirements. Irrigation and drainage paper, 24.
56. Doorenbos, J., & Kassam, A. H. (1979). Yield response to water. Irrigation and drainage paper, 33, 257.
57. Dunkle, R. V. (1940). Heat meters. Bulletin of the American Meteorological Society, 21(3), 116-117.
58. Dunne, T., & Black, R. D. (1970). An experimental investigation of runoff production in permeable soils. Water Resources Research, 6(2),

478-490.

59. Enz, J. W., Brun, L. J., & Larsen, J. K. (1988). Evaporation and energy balance for bare and stubble covered soil. Agricultural and forest meteorology, 43(1), 59-70.
60. Fajer, E. D., & Bazzaz, F. A. (1992). Is carbon dioxide a 'good' greenhouse gas?: Effects of increasing carbon dioxide on ecological systems. Global Environmental Change, 2(4), 301-310.
61. Falckenberg, G. (1930). ApparaturzurBestimmung des momentanennächtlichenWärmeaustauscheszwischenErde und Luft. Meteorol. Zeitschrift, 47, 154-156.
62. Franzke, C. L., Barbosa, S., Blender, R., Fredriksen, H. B., Laepple, T., Lambert, F., ... & Yuan, N. (2020). The structure of climate variability across scales. Reviews of Geophysics, 58(2), e2019RG000657.
63. FROST, W. (1973). Review of data and prediction techniques for wind profiles around manmade surface obstructions (summary of analytical models and calculational procedures). AGARD Flight in Turbulence 18 p(SEE N 74-17720 09-02).
64. Fuchs, M. (1986). Heat flux. Methods of Soil Analysis: Part 1 Physical and Mineralogical Methods, 5, 957-968.
65. Fuchs, M. (1986). Heat flux.--p. 957-968 (No. Colección general/631.41 M592m v. 1). En: Methods of soil analysis: part 1; physical and mineralogical methods.--Winsconsin, US: American Society of Agronomy, 1986.
66. Fuchs, M., & Tanner, C. B. (1968). Calibration and field test of soil heat flux plates. Soil Science Society of America Journal, 32(3), 326-328.
67. Fuchs, M., &Hadas, A. (1972). The heat flux density in a non-homogeneous bare loessial soil. Boundary-layer meteorology, 3(2), 191-200.
68. Gabrielsen, E. K. (1960). Light wavelengths and photosynthesis. Handbook of Plant Physiology. Springer, Berlin.
69. Gandemer, J., & Guyot, A. (1981). La protection contre le vent: aerodynamique des brise-vent et conseils pratiques. Centre scientifique et technique du batiment.
70. Gandin, L. S. (1970). The planning of meteorological station networks.
71. Gatenby, R. M. (1977). Conduction of heat from sheep to ground. Agricultural meteorology, 18(5), 387-400.

72. Gausman, H. W., Gerbermann, A. H., Wiegand, C. L., Leamer, R. W., Rodriguez, R. R., & Noriega, J. R. (1975). Reflectance differences between crop residues and bare soils. Soil Science Society of America Journal, 39(4), 752-755.
73. Gee, G. W., Or, D., Dane, J. H., &Topp, G. C. (2002). Methods of soil analysis. Part 4. Physical methods. Methods of soil analysis.
74. Gee, G. W., Ward, A. L., Caldwell, T. G., & Ritter, J. C. (2002). A vadose zone water fluxmeter with divergence control. Water Resources Research, 38(8), 16-1.
75. Geiger, R. (1966). The climate near the ground Harvard Univ. Press, Cambridge, Mass.
76. Gerber, E. P., Butler, A., Calvo, N., Charlton-Perez, A., Giorgetta, M., Manzini, E., ... & Watanabe, S. (2012). Assessing and understanding the impact of stratospheric dynamics and variability on the Earth system. Bulletin of the American Meteorological Society, 93(6), 845-859.
77. Gilgen, H., & Ohmura, A. (1999). The global energy balance archive. Bulletin of the American Meteorological Society, 80(5), 831-850.
78. Gillespie, T. J., & Barr, A. (1984). Adaptation of a dew estimation scheme to a new crop and site. Agricultural and forest meteorology, 31(3-4), 289-295.
79. Gimeno, L., Stohl, A., Trigo, R. M., Dominguez, F., Yoshimura, K., Yu, L., ... & Nieto, R. (2012). Oceanic and terrestrial sources of continental precipitation. Reviews of Geophysics, 50(4).
80. Göbel, W. (1995). Adding stochasticity to GIS: stochastic regional models. FAO Agrometeorology Series Working Paper (FAO).
81. Gommes, R. (1998). List of variables used for crop modelling purposes by the Space Applications Institute/MARS. Rijks, D., Terres, JM, Vossen, P.(Eds.), 313-326.
82. Gommes, R. (1999). Extreme Agrometeorological Events, Discussion leading to the terms of reference for a database. WMO CagM-12 WG on extreme agrometeorological events.
83. Görlich, P. (1951). The photocells.
84. Grossman, R. B., &Reinsch, T. G. (2002). 2.1 Bulk density and linear extensibility. Methods of soil analysis: Part 4 physical methods, 5, 201-228.
85. Haeffelin, M., Kato, S., Smith, A. M., Rutledge, C. K., Charlock, T. P., & Mahan, J. R. (2001). Determination of the thermal offset of the

Eppley precision spectral pyranometer. Applied optics, 40(4), 472-484.

86. Ham, J. M., &Kluitenberg, G. J. (1993). Positional variation in the soil energy balance beneath a row-crop canopy. Agricultural and forest meteorology, 63(1-2), 73-92.

87. Ham, J. M., Heilman, J. L., &Lascano, R. J. (1991). Soil and canopy energy balances of a row crop at partial cover. Agronomy Journal, 83(4), 744-753.

88. Hamdan, Ahmad & Mustapha, Faizal & Nizam, Ahmad & Abdul Latif, Mohd Faruq. (2014). Optimization of Accelerated Ducted Design in Wind Turbine System Using Computational Fluid Dynamics (CFD). Australian journal of basic and applied sciences. 8. 264-268.

89. Hanks, R. J., & Jacobs, H. S. (1971). Comparison of the calorimetric and flux meter measurements of soil heat flow. Soil Science Society of America Journal, 35(5), 671-674.

90. Hanks, R. J., & Jacobs, H. S. (1971). Comparison of the calorimetric and flux meter measurements of soil heat flow. Soil Science Society of America Journal, 35(5), 671-674.

91. Harazono, Y., Kim, J., Miyata, A., Choi, T., Yun, J. I., & Kim, J. W. (1998). Measurement of energy budget components during the International Rice Experiment (IREX) in Japan. Hydrological processes, 12(13-14), 2081-2092.

92. Hartmann, D. L. (2015). Global physical climatology (Vol. 103). Newnes.

93. Heilman, J. L., McInnes, K. J., Savage, M. J., Gesch, R. W., &Lascano, R. J. (1994). Soil and canopy energy balances in a west Texas vineyard. Agricultural and Forest Meteorology, 71(1-2), 99-114.

94. Hernandez-Ramirez, G., Hatfield, J. L., Prueger, J. H., & Sauer, T. J. (2010). Energy balance and turbulent flux partitioning in a corn–soybean rotation in the Midwestern US. Theoretical and Applied Climatology, 100, 79-92.

95. Hooghart, J. C. (1987). Evaporation and weather. TNO Committe on Hydrological Research.

96. Hoppe, P. (2000). Challenges for biometeorology on the turn of the century. Biometeorology and Urban Climatology at the Turn of the Millennium, 383-386.

97. Horton, R., &Wierenga, P. J. (1983). Estimating the soil heat flux from observations of soil temperature near the surface. Soil Science Society

of America Journal, 47(1), 14-20.

98. Hossen, M. S., Mano, M., Miyata, A., Baten, M. A., &Hiyama, T. (2012). Surface energy partitioning and evapotranspiration over a double - cropping paddy field in Bangladesh. Hydrological Processes, 26(9), 1311-1320.
99. Hossen, M. S., Mano, M., Miyata, A., Baten, M. A., &Hiyama, T. (2012). Surface energy partitioning and evapotranspiration over a double - cropping paddy field in Bangladesh. Hydrological Processes, 26(9), 1311-1320.
100. Hubbard, R. K., & Sheridan, J. M. (1983). Water and nitrate - nitrogen losses from a small, upland, coastal plain watershed (Vol. 12, No. 2, pp. 291-295). American Society of Agronomy, Crop Science Society of America, and Soil Science Society of America.
101. Huschke, R. E. (1959). Glossary of meteorology.
102. Hutte, P. (1968). Experiments on windflow and wind damage in Germany: site and susceptibility of Spruce forests to storm damage. Forestry, (Suppl.), 20-7.
103. Hyndman, B. (2020). Heating, ventilation, and air conditioning. In Clinical engineering handbook (pp. 662-666). Academic Press.
104. Idso, S. B. (1972). Calibration of soil heat flux plates by a radiation technique. Agricultural Meteorology, 10, 467-471.
105. IRRI. 2009. Rice ecosystems. International Rice Research Institute, Los Baños, The Philippines. http://www.ipni.net/ppiweb/filelib.nsf/0/6191D544DF714DEF48257074002E78E6/ $file/Rice%20HB%20p2-5.pdf (verified 15 June 2018).
106. Jackson, R. D., Idso, S. B., Reginato, R. J., & Pinter Jr, P. J. (1981). Canopy temperature as a crop water stress indicator. Water resources research, 17(4), 1133-1138.
107. Jacobs, A. F. G. (1985). Turbulence around a thin solid fence. Agricultural and forest meteorology, 34(4), 315-321.
108. Jacobs, A. F., &Nieveen, J. P. (1995). Formation of dew and the drying process within crop canopies. Meteorological Applications, 2(3), 249-256.
109. Jacobs, A. F., Heusinkveld, B. G., &Holtslag, A. A. (2011). Long-term record and analysis of soil temperatures and soil heat fluxes in a grassland area, The Netherlands. Agricultural and Forest Meteorology, 151(7), 774-780.

110. Jensen, M. E., Burman, R. D., & Allen, R. G. (1990). Evapotranspiration and irrigation water requirements. ASCE.
111. Jessen, W., Wilbert, S., Zarzalejo, L. F., Ramirez-Santigosa, L., Valenzuela, R., Liria, J., ... &Hanrieder, N. (2019). Calibration Procedures for Rotating ShadowbandIrradiometers.
112. Jia, Z., Zhang, W., & Huang, Y. (2010). Analysis of energy flux in rice paddy in the Sanjiang Plain. ZhongguoShengtaiNongyeXuebao/ Chinese Journal of Eco-Agriculture, 18(4), 820-826.
113. Jo, Y. H., Lee, C. H., &Yoo, J. H. (2013). Study on applicability of passive infrared thermography analysis for blistering detection of stone cultural heritage. Journal of Conservation Science, 29(1), 55-67.
114. Kaempfert, W. (1942). Solar radiation on plane, wall, and slope. Reich Office for Weather Service: Wiss. Abh. B, IX.
115. Keane, T., & Collins, J. F. (1986). Climate, weather and Irish agriculture. Dublin: Agmet.
116. Kimball, B. A., & Jackson, R. D. (1975). Soil heat flux determination: A null-alignment method. Agricultural Meteorology, 15(1), 1-9.
117. Kimball, B. A., LaMorte, R. L., Pinter Jr, P. J., Wall, G. W., Hunsaker, D. J., Adamsen, F. J., ... & Brooks, T. J. (1999). Free-air CO2 enrichment and soil nitrogen effects on energy balance and evapotranspiration of wheat. Water resources research, 35(4), 1179-1190.
118. Kluitenberg, G. J. (2002). 5.2 Heat capacity and specific heat. Methods of Soil Analysis: Part 4 Physical Methods, 5, 1201-1208.
119. Kluitenberg, G. J., Ham, J. M., & Bristow, K. L. (1993). Error analysis of the heat pulse method for measuring soil volumetric heat capacity. Soil Science Society of America Journal, 57(6), 1444-1451.
120. Konstantinov, A. R., Sakaly, L. I., &Diagot, L. S. (1974). THE SIGNIFICANCE OF THE COMPONENTS OF HEAT AND WATER BALANCES IN THE FORMATION OF MICRO-AND LOCAL CLIMATE. In Physical and Dynamic Climatology: Proceedings of the Symposium on Physical and Dynamic Climatology (No. 347, p. 99). Gidrometeoizdat.
121. Kool, D., Heitman, J. L., Lazarovitch, N., Agam, N., Sauer, T. J., & Ben-Gal, A. (2016). In situ thermistor calibration for improved measurement of soil temperature gradients. Soil Science Society of America Journal, 80(6), 1514-1519.
122. Kruizinga, S., &Yperlaan, G. J. (1978). Spatial interpolation of daily

totals of rainfall. Journal of Hydrology, 36(1-2), 65-73.

123. Kumar, K. K., & Rao, V. U. (1984). Significance of soil heat flux in the estimation of potential evapotranspiration. Indian Journal of Ecology.

124. Kustas, W. P., & Daughtry, C. S. (1990). Estimation of the soil heat flux/net radiation ratio from spectral data. Agricultural and Forest Meteorology, 49(3), 205-223.

125. Kustas, W. P., Daughtry, C. S., & Van Oevelen, P. J. (1993). Analytical treatment of the relationships between soil heat flux/net radiation ratio and vegetation indices. Remote sensing of environment, 46(3), 319-330.

126. Kustas, W. P., Prueger, J. H., Hatfield, J. L., Ramalingam, K., &Hipps, L. E. (2000). Variability in soil heat flux from a mesquite dune site. Agricultural and Forest Meteorology, 103(3), 249-264.

127. Kustas, W. P., Prueger, J. H., Hatfield, J. L., Ramalingam, K., &Hipps, L. E. (2000). Variability in soil heat flux from a mesquite dune site. Agricultural and Forest Meteorology, 103(3), 249-264.

128. Lange, B. (1939). The photocells and their application, Vols. I, II.

129. Lemon, E. R., Shinn, J. H., & Stoller, J. H. (1963). Experimental determination of the energy balance. The energy budget at the earth's surface, Part I. Prod. Res. Rep, (71).

130. Lettau, H., & Davidson, B. (Eds.). (1957). Exploring the Atmosphere's First Mile: Site Description and Data Tabulation. Pergamon Press.

131. Lewis, J. M. (1995). The story behind the Bowen ratio. Bulletin of the American Meteorological Society, 76(12), 2433-2444.

132. Linke, F., & Baur, F. (1953). Meteorological pocketbook.

133. Lovegrove, K., & Stein, W. (Eds.). (2012). Concentrating solar power technology: principles, developments and applications.

134. Lowry, W. P. (1967). Weather and Life Academic Press. New York, New York, USA.

135. Mahrt, L., Heald, R. C., Lenschow, D. H., Stankov, B. B., &Troen, I. B. (1979). An observational study of the structure of the nocturnal boundary layer. Boundary-Layer Meteorology, 17, 247-264.

136. Makkink, G. F. (1957). Ekzameno de la formulo de Penman. Neth. J. Agric. Sci, 5, 290-305.

137. Malek, E. (1993). Rapid changes of the surface soil heat flux and its effects on the estimation of evapotranspiration. Journal of Hydrology,

142(1-4), 89-97.

138. Maruyama, A., &Kuwagata, T. (2010). Coupling land surface and crop growth models to estimate the effects of changes in the growing season on energy balance and water use of rice paddies. Agricultural and Forest Meteorology, 150(7-8), 919-930.
139. Maselli, F., Petkov, L., &Maracchi, G. (1998). Extension of climate parameters over the land surface by the use of NOAA-AVHRR and ancillary data. Photogrammetric engineering and remote sensing, 64, 199-206.
140. Masseroni, D., Facchi, A., Romani, M., Chiaradia, E. A., Gharsallah, O., & Gandolfi, C. (2015). Surface energy flux measurements in a flooded and an aerobic rice field using a single eddy-covariance system. Paddy and Water Environment, 13, 405-424.
141. Massman, W. J. (1992). Correcting errors associated with soil heat flux measurements and estimating soil thermal properties from soil temperature and heat flux plate data. Agricultural and Forest Meteorology, 59(3-4), 249-266.
142. Massman, W. J. (1993). Errors associated with the combination method for estimating soil heat flux. Soil Science Society of America Journal, 57(5), 1198-1202.
143. Mayocchi, C. L., & Bristow, K. L. (1995). Soil surface heat flux: some general questions and comments on measurements. Agricultural and forest meteorology, 75(1-3), 43-50.
144. McCree, K. J. (1973). The measurement of photosynthetically active radiation. Solar energy, 15(1), 83-87.
145. McNaughton, K. G., & Jarvis, P. G. (1984). Using the Penman-Monteith equation predictively. Agricultural Water Management, 8(1-3), 263-278.
146. Mironova, Z. F. (1973). Albedo of earth's surface and clouds. Radiation Characteristics of the Atmosphere and the Earth's Surface (RadiatsionnyeKharaketeristikiAtmosfery i ZemnoiPoverkhnosti). TT 71-58003--NASA TTF-678--Category 13, 192.
147. Miyata, A., Leuning, R., Denmead, O. T., Kim, J., &Harazono, Y. (2000). Carbon dioxide and methane fluxes from an intermittently flooded paddy field. Agricultural and Forest Meteorology, 102(4), 287-303.
148. Mogensen, V. O. (1970). The calibration factor of heat flux meters

in relation to the thermal conductivity of the surrounding medium. Agricultural Meteorology, 7, 401-410.

149. Molga, M. (1962). Agricultural Meteorology. Part II Cutline of Agrometeological Problems (No. 551.5/M719).
150. Monteith, J. L. (1957). Dew. Quarterly Journal of the Royal Meteorological Society, 83(357), 322-341.
151. Monteith, J. L. (1965). Evaporation and environment. In Symposia of the society for experimental biology (Vol. 19, pp. 205-234). Cambridge University Press (CUP) Cambridge.
152. Monteith, J. L. (1965). The state and movement of water in living organisms. In 19th Symposia of the Society for Experimental Biology. Cambridge University Press, London, 1965 (pp. 205-234).
153. Mount, L. E. (1979). Adaptation to thermal environment. Man and his productive animals. Edward Arnold (Publishers) Ltd..
154. Munn, R. E. (1966). Descriptive Micrometeorology (Supplement 1).
155. Newman, J. E. (1974). Applying meteorology to agriculture. Agricultural Meteorology, 13(1), 1-3.
156. Nicolet, M., &Dogniaux, R. (1951). Study of the solar global radiation. Mem. Inst. R. Meteorol. Belg. XLV, II.
157. Ochsner, T. E., Horton, R., & Ren, T. (2001). A new perspective on soil thermal properties. Soil science society of America Journal, 65(6), 1641-1647.
158. Ochsner, T. E., Sauer, T. J., & Horton, R. (2007). Soil heat storage measurements in energy balance studies. Agronomy Journal, 99(1), 311-319.
159. Ogallo, L. A., Boulahya, M. S., & Keane, T. (2000). Applications of seasonal to interannual climate prediction in agricultural planning and operations. Agricultural and Forest Meteorology, 103(1-2), 159-166.
160. Oke, T. R. (1987). Boundary layer climates, 2nd (ed.) Routledge. London and Wiley, New York.
161. Oliver, H. R. (1992). Studies of surface energy balance of sloping terrain. International journal of climatology, 12(1), 55-68.
162. Oliver, H. R. (1992). Studies of surface energy balance of sloping terrain. International journal of climatology, 12(1), 55-68.
163. Oliver, J. E., &Hidore, J. J. (2002). Climatology. An atmospheric science. Prentice Hall.

164. Paoli, C., Voyant, C., Muselli, M., &Nivet, M. L. (2009). Solar radiation forecasting using ad-hoc time series preprocessing and neural networks. In Emerging Intelligent Computing Technology and Applications: 5th International Conference on Intelligent Computing, ICIC 2009, Ulsan, South Korea, September 16-19, 2009. Proceedings 5 (pp. 898-907). Springer Berlin Heidelberg.

165. Parkin, G.W. (2008). Water Budget In Soil. In: Chesworth, W. (eds) Encyclopedia of Soil Science. Encyclopedia of Earth Sciences Series. Springer, Dordrecht. https://doi.org/10.1007/978-1-4020-3995-9_622

166. Pedro Jr, M. J., & Gillespie, T. J. (1981). Estimating dew duration. II. Utilizing standard weather station data. Agricultural Meteorology, 25, 297-310.

167. Penman, H. L. (1948). Natural evaporation from open water, bare soil and grass. Proceedings of the Royal Society of London. Series A. Mathematical and Physical Sciences, 193(1032), 120-145.

168. PENMAN, H. T. (1951). Some physical aspects of assimilation and transpiration. In Symposia Soc. Exptl. Biol. (Vol. 5, pp. 115-129).

169. Perry, A. E., Schofield, W. H., & Joubert, P. N. (1969). Rough wall turbulent boundary layers. Journal of Fluid Mechanics, 37(2), 383-413.

170. Petkov, L., Pieri, M., Maselli, F., &Maracchi, G. (1996). Study and modelling of temperature spatial variability by NOAA-AVHRR thermal imagery. ISPRS journal of photogrammetry and remote sensing, 51(3), 127-136.

171. Philip, J. R. (1961). The theory of heat flux meters. Journal of Geophysical Research, 66(2), 571-579.

172. Philip, J. R., & De Vries, D. D. (1957). Moisture movement in porous materials under temperature gradients. Eos, Transactions American Geophysical Union, 38(2), 222-232.

173. Pikul Jr, J. L., &Allmaras, R. R. (1984). A field comparison of null-aligned and mechanistic soil heat flux. Soil Science Society of America Journal, 48(6), 1207-1214.

174. Portman, D. J. (1958). Conductivity and length relationships in heat-flow transducer performance. Eos, Transactions American Geophysical Union, 39(6), 1089-1094.

175. Priestley, C. H. B., & Taylor, R. J. (1972). On the assessment of surface heat flux and evaporation using large-scale parameters. Monthly weather review, 100(2), 81-92.

176. Priestley, C. H. B., & Taylor, R. J. (1972). On the assessment of surface heat flux and evaporation using large-scale parameters. Monthly weather review, 100(2), 81-92.

177. Ramegowda, V., & Senthil-Kumar, M. (2015). The interactive effects of simultaneous biotic and abiotic stresses on plants: mechanistic understanding from drought and pathogen combination. Journal of plant physiology, 176, 47-54.

178. Rao, G. R., Ramamohan, R. V., & Rao, B. R. (1977). Soil heat flux studies in the bare red laterite soil during winter and summer months at Bangalore [India]. Annals of Arid Zone (India).

179. Raupach, M., & Thom, A. S. (1981). Turbulence in and above plant canopies. Annual Review of Fluid Mechanics, 13(1), 97-129.

180. Richard, G., & Cellier, P. (1998). Effect of tillage on bare soil energy balance and thermal regime: an experimental study. Agronomie, 18(3), 163-181.

181. Rijks, D., &Baradas, M. W. (2000). The clients for agrometeorological information. Agricultural and Forest Meteorology, 103(1-2), 27-42.

182. Rijtema, P. E. (1965). An analysis of actual evapotranspiration (Doctoral dissertation, Pudoc).

183. Ripley, E. A., &Redmann, R. E. (1976). Grassland. Vegetation and the Atmosphere, Vol. 2, JL Monteith, Ed. Academic Press, 351, 396.

184. Robin, P., Cellier, P., & Richard, G. (1997). Theoretical and field comparison of two types of soil heat fluxmeter. Soil technology, 10(3), 185-206.

185. RS, S. (1978). Environmental aerodynamics. New York: John Wiley & Sons.

186. Rudra, R. P., Dickinson, W. T., Whiteley, H. R., Gayner, J. D., & Wall, G. J. (2000). SELECTION OF APPROPRIATE EVAPORATION ESTIMATION TECHNIQUE FOR CONTINUOUS MODELING 1. JAWRA Journal of the American Water Resources Association, 36(3), 585-594.

187. Rutter, N. (1968). Geomorphic and tree shelter in relation to surface wind conditions, weather, time of day and season. Agricultural Meteorology, 5(5), 319-334.

188. Sadek, M. F., Shahin, M. M., &Stigter, C. J. (1997). Evaporation from the reservoir of the High Aswan Dam, Egypt: A new comparison of relevant methods with limited data. Theoretical and Applied

Climatology, 56, 57-66.

189. Sánchez, B., Rasmussen, A., & Porter, J. R. (2014). Temperatures and the growth and development of maize and rice: a review. Global change biology, 20(2), 408-417.
190. Sarma, A.A.L.N. (1987). Scales of climate . In: Climatology. Encyclopedia of Earth Science. Springer, Boston, MA. https://doi.org/10.1007/0-387-30749-4_153
191. Sauer, T. J. (2002). Heat flux density. p. 1233–1248. JH Dane and GC Topp (ed.) Methods of soil analysis. Part 4. SSSA Book Ser. 5. SSSA, Madison, WI. Heat flux density. p. 1233–1248. In JH Dane and GC Topp (ed.) Methods of soil analysis. Part 4. SSSA Book Ser. 5. SSSA, Madison, WI.
192. Sauer, T. J., & Horton, R. (2005). Soil heat flux. Micrometeorology in agricultural systems, 47, 131-154.
193. Sauer, T. J., & Horton, R. (2005). Soil heat flux. Micrometeorology in agricultural systems, 47, 131-154.
194. Sauer, T. J., & Norman, J. M. (1995). Simulated canopy microclimate using estimated below-canopy soil surface transfer coefficients. Agricultural and forest meteorology, 75(1-3), 135-160.
195. Sauer, T. J., & Peng, X. (2020). Soil temperature and heat flux. Agroclimatology: Linking Agriculture to Climate, 60, 73-93.
196. Sauer, T. J., Hatfield, J. L., &Prueger, J. H. (1997). Over-winter changes in radiant energy exchange of a corn residue-covered surface. Agricultural and forest meteorology, 85(3-4), 279-287.
197. Sauer, T. J., Hatfield, J. L., Prueger, J. H., & Norman, J. M. (1998). Surface energy balance of a corn residue-covered field. Agricultural and Forest Meteorology, 89(3-4), 155-168.
198. Sauer, T. J., Norman, J. M., Tanner, C. B., & Wilson, T. B. (1995). Measurement of heat and vapor transfer coefficients at the soil surface beneath a maize canopy using source plates. Agricultural and Forest Meteorology, 75(1-3), 161-189.
199. Scaëtta, H. (1935). Les avalanches d'air dans les Alpes et dans les hautesmontagnes de l'Afrique centrale. Ciel et Terre, 51, 79.
200. Scaetta, H. (1935). Terminologieclimatique, bioclimatique et microclimatique. La Mét, 11, 342-347.
201. Schulze, R. (1949). Effect-related radiation in meteorology. Ber. d. DWD, Special Ed. of Meteorol. Rundsch.

202. Schulze, R. (1953). On a radiation measuring device with UV transmitting wind protection cap at the meteorological Observatory Hamburg. Geofis. Pura Appl, 24.

203. Schulze, R. (1970). Radiation climate of the earth. Strahlentherapie, 130(4), 595-599.

204. Seemann, J. (1965). A contribution of the carbon dioxide problem in greenhouses. Gartenbauwissenschaft *4*.

205. Seemann, J., Chirkov, Y. I., Lomas, J., &Primault, B. (2012). Agrometeorology. Springer Science & Business Media.

206. Seemann, J., Chirkov, Y. I., Lomas, J., Primault, B., & Seemann, J. (1979). Greenhouse climate. Agrometeorology, 165-178.

207. Seginer, I. D. O. (1975). Atmospheric-stability effect on windbreak shelter and drag. Boundary-Layer Meteorology, 8, 383-400.

208. Seguin, B. (1991, November). Utilisation combinée de donnéessatellitaires et de modèlesagrométéorologiques. In Conference.

209. Sellers, W. D. (1969). A global climatic model based on the energy balance of the earth-atmosphere system. Journal of Applied Meteorology and Climatology, 8(3), 392-400.

210. Sellers, W. D., & Climatology, P. (1965). Illinois.

211. Sharratt, B. S., Campbell, G. S., & Glenn, D. M. (1992). Soil heat flux estimation based on the finite-difference form of the transient heat flow equation. Agricultural and forest meteorology, 61(1-2), 95-111.

212. Shenoy, V., Tripathi, P., Mahadik, A., Nalawade, P., & Mahajan, A. (2018). Devices Used for Measuring Solar Radiation–A Review. In International Conference on Innovative and Advanced Technologies in Engineering. IOSR Journal of Engineering (IOSRJEN) www. iosrjen. org ISSN (e) (pp. 2250-3021).

213. Slatyer, R. 0. and McIlroy. IC:" (1961). Practical microclimatology with special reference to the water factor in soil-plant-atmosphere relationship" CSIRO, Melbourse.

214. Slatyer, R. O. (1969). Physiological significance of internal water relations to crop yield. Physiological aspects of crop yield, 53-83.

215. Staley, R. C., Gerhardt, J. R., Lettau, H. H., & Davidson, B. (1957). Soil heat flux measurements. Exploring the atmosphere's first mile. Pergamon Press, Elmsford, NY, 58-63.

216. Stricker, H., &Brutsaert, W. (1978). Actual evapotranspiration over

a summer period in the "Hupsel catchment". Journal of Hydrology, 39(1-2), 139-157.

217. Stull, R. B. (1988). An introduction to boundary layer meteorology (Vol. 13). Springer Science & Business Media.
218. Sturman, A. P. (2001). Local and regional weather and climate.
219. Suzuki, N., Rivero, R. M., Shulaev, V., Blumwald, E., &Mittler, R. (2014). Abiotic and biotic stress combinations. New Phytologist, 203(1), 32-43.
220. Tamai, K., Abe, T., Araki, M., & Ito, H. (1998). Radiation budget, soil heat flux and latent heat flux at the forest floor in warm, temperate mixed forest. Hydrological processes, 12(13 - 14), 2105-2114.
221. Tanner, C. B. (1963). Basic instrumentation and measurements for plant environment and micrometeorology. Soils Bull.
222. Tennekes, H. (1973). The logarithmic wind profile. Journal of Atmospheric Sciences, 30(2), 234-238.
223. Thornthwaite, C. W. (1953). Topoclimatology. In Proceeding Toronto Meteorological, Conference (pp. 227-232). American Meteorological Society and Royal Meteorological Society.
224. Thornthwaite, C. W. (1957). Instructions and tables for computing potential evapotranspiration and the water balance. Publications on Climatology, 10, 185-310.
225. Tindall, J. A., Kunkel, J. R., & Anderson, D. E. (1999). Unsaturated zone hydrology for scientists and engineers (Vol. 4). Upper Saddle River, NJ: Prentice Hall.
226. Topp, G. C., &Ferre, T. P. A. (2002). Water content, In, Methods of Soil Analysis. Part 4.(Ed. JH Dane and GC Topp), SSSA Book Series No. 5. Soil Science Society of America, Madison WI.
227. Topp, G. C., Davis, J. L., & Annan, A. P. (1980). Electromagnetic determination of soil water content: Measurements in coaxial transmission lines. Water resources research, 16(3), 574-582.
228. Topp, G. C., Davis, J. L., & Annan, A. P. (1980). Electromagnetic determination of soil water content: Measurements in coaxial transmission lines. Water resources research, 16(3), 574-582.
229. Topp, G. C., Ferré, P. A., & Dane, J. H. (2002). Water content. Methods of soil analysis. Part, 4, 417-545.
230. Tsai, J. L., Tsuang, B. J., Lu, P. S., Yao, M. H., & Shen, Y. (2007).

Surface energy components and land characteristics of a rice paddy. Journal of applied meteorology and climatology, 46(11), 1879-1900.

231. Tuzet, A., Castell, J. F., Perrier, A., &Zurfluh, O. (1997). Flux heterogeneity and evapotranspiration partitioning in a sparse canopy: the fallow savanna. Journal of Hydrology, 188, 482-493.

232. Twine, T. E., Kustas, W. P., Norman, J. M., Cook, D. R., Houser, P., Meyers, T. P., ... &Wesely, M. L. (2000). Correcting eddy-covariance flux underestimates over a grassland. Agricultural and forest meteorology, 103(3), 279-300.

233. Unger, P. W. (1988). Residue management effects on soil temperature. Soil Science Society of America Journal, 52(6), 1777-1782.

234. United States. Department of Agriculture. Economics, Statistics Service, United States. Foreign Agricultural Service, United States. World Food, Agricultural Outlook, & Situation Board. (2002). World agricultural supply and demand estimates. The Department.

235. Utaaker, K. (1974). Physics of local and microclimate. From: Physical and Dynamic Climatology, WMO Publication, (347).

236. Van Dam, A. M., &Leffelaar, P. A. (1998). Root, soil water and nitrogen dynamics in a catch crop-soil system in the Wageningen Rhizolab. Netherlands Journal of agricultural science, 267-284.

237. Van Loon, W. K. P., Bastings, H. M. H., & Moors, E. J. (1998). Calibration of soil heat flux sensors. Agricultural and Forest Meteorology, 92(1), 1-8.

238. Van Wijk, W. R. (1967). New method for measuring heat flux density at the surface of soils or of other bodies. Nature, 213, 214-214.

239. VANARSDE. EP. (1965, January). Micrometeorology and plant disease epidemiology. In Phytopathology (Vol. 55, No. 9, pp. 945-+). 3340 PILOT KNOB ROAD, ST PAUL, MN 55121 USA: AMER PHYTOPATHOLOGICAL SOC.

240. Vehrencamp, J. E. (1953). Experimental investigation of heat transfer at an air - Earth interface. Eos, Transactions American Geophysical Union, 34(1), 22-30.

241. Ventskevich, G. Z., & Rudenko, A. I. (1961). Agrometeorology (No. 631.9 VEN).

242. Verma, S. B. (1990). Micrometeorological methods for measuring surface fluxes of mass and energy. Remote sensing reviews, 5(1), 99-115.

243. Wang, J., & Bras, R. L. (1999). Ground heat flux estimated from surface soil temperature. Journal of hydrology, 216(3-4), 214-226.

244. Waring, R. H., & Running, S. W. (2007). Forest ecosystems: analysis at multiple scales. Third Edition Elsevier.

245. Watts, D. B., Kanemasu, E. T., & Tanner, C. B. (1990). Modified heat-meter method for determining soil heat flux. Agricultural and forest meteorology, 49(4), 311-330.

246. Welch, S. M., Kluitenberg, G. J., & Bristow, K. L. (1996). Rapid numerical estimation of soil thermal properties for a broad class of heat-pulse emitter geometries. Measurement Science and Technology, 7(6), 932.

247. Whipkey, R. Z. (1965). Subsurface stormflow from forested slopes. Hydrological Sciences Journal, 10(2), 74-85.

248. Wieringa, J. (1980). Representativeness of wind observations at airports. Bulletin of the American Meteorological Society, 61(9), 962-971.

249. Wieringa, J. (1996). Does representative wind information exist?. Journal of wind engineering and Industrial Aerodynamics, 65(1-3), 1-12.

250. Wieringa, J., & Lomas, J. (2001). Lecture notes for training agricultural meteorological personnel. Secretariat of the World Meteorological Organization.

251. Wieringa, J., & Lomas, J. (2001). Lecture notes for training agricultural meteorological personnel. Secretariat of the World Meteorological Organization.

252. Wilson, J. W., &Brandes, E. A. (1979). Radar measurement of rainfall—A summary. Bulletin of the American Meteorological Society, 60(9), 1048-1060.

253. Wmo. (1981). Guide To Agricultural Meteorological Practices. Selbstverl..

254. Wollny, E. (1878). Untersuchungenüber den Einfluß der Exposition auf die Erwärmung des Bodens. Forschungen auf dem gebiete der Agricultur-physik (Heidelberg), 1, 263-294.

255. Woodward, F. I., & Woodward, F. I. (1987). Climate and plant distribution. Cambridge University Press.

256. Wörner, H. (1955). The stability of selenium photocells. Z. Meteorol, 9.

257. Yamori, W., Hikosaka, K., & Way, D. A. (2014). Temperature response of photosynthesis in C 3, C 4, and CAM plants: temperature acclimation and temperature adaptation. Photosynthesis research, 119, 101-117.

258. Yoshino, M. M. (1975). Climate in a small area: an introduction to local meteorology (Vol. 549). Tokyo: University of Tokyo press.

INDEX

M

N

O

P

W

Y

Z